Keys to Sacred Science

بسم الله الرحمن الرحيم

Keys to Sacred Science:
Geometry and Numerology in Islam

by Suleiman Sami Joukhadar

FONS VITAE

First published in 2018 by
Fons Vitae
49 Mockingbird Valley Drive
Louisville, KY 40207
http://www.fonsvitae.com
Email: fonsvitaeky@aol.com

Library of Congress Control Number: 2018958893
ISBN 978-1891785-771

We would like to thank Kaoutar El-Mernissi and Jane Fatima Casewit for their thorough translation work. Our thanks go also to all the editing team members for their helpful contributions.

Printed in Morocco and the United States of America

المطبعة والوراقة الوطنية
IMPRIMERIE PAPETERIE EL WATANYA
زنقة أبو عبيدة، الحي المحمدي، الداوديات - مراكش
RUE ABOU OUBAIDA, CITE MOHAMMADIA, DAOUDIAT MARRAKECH
TEL.: 05 24 30 37 74 LG / 05 24 30 25 91 - FAX: 05 24 30 49 23
iwatanya@gmail.com www.elwatanya.ma

Contents

Author's Biography . . . 11
Preamble . . . 13
1. An Overview of the Structure and Methodology Used . . . 14
2. Review of Some Cultural and Educational Concepts . . . 16
3. Do the Written Sources Define the Islamic Heritage? . . . 22
4. The Challenges of Communicating Knowledge or Thought in Words . . . 27
5. Prelude to the General Introduction to Sacred Knowledge . . . 30
General Introduction to Sacred Knowledge . . . 35
The Story of Moses and al-Khiḍr (Peace and Blessings Be Upon Them) . . . 37
1. And When (*Wa Idh*) . . . 39
2. Knowledge . . . 41
3. Adam and Knowledge . . . 45
4. Iblīs and Ignorance . . . 47
5. Iblīs's Errors . . . 55
6. The Human Drama . . . 56
7. The Escape . . . 57
8. Our Master Moses . . . 58
9. The Young Man (*al-Fatā*) . . . 59
10. The Fish (*al-Ḥūt*) . . . 60
11. Forgetfulness . . . 62
12. The Wise One . . . 63
13. Patience . . . 64
14. Silence, Gesture and Allusion . . . 65
15. So They Went On . . . 67
16. The Final Outcome . . . 69
17. Knowledge . . . 71
Summary of Chapters . . . 73
Introduction to the Treatise . . . 77
1. The Wafaq Square . . . 77
2. The Triangle . . . 78
3. The Treatise . . . 79
4. Self-effacement . . . 81
5. The Point and the Circle . . . 82
6. Allusion . . . 83

Treatise on the 3×3 Wafaq Square 85
The 3×3 Wafaq Square. 86
Primary, Basic, and Grand Rules 96
Primary Rules 97
Basic Rules. 97
,... . . . 97
Grand Rules 97
Definitions 133
The Number *Pi* 133
The Tetraktys or the Triangle of the Gnostics 134
The Tarot 135
The *Gestalt* Theory 136
Ḥisāb al-Jumal or the *Abjad* System 137
The Golden Ratio: φ. 138
The Prime Numbers 138

في مجال المعرفة
المركزي

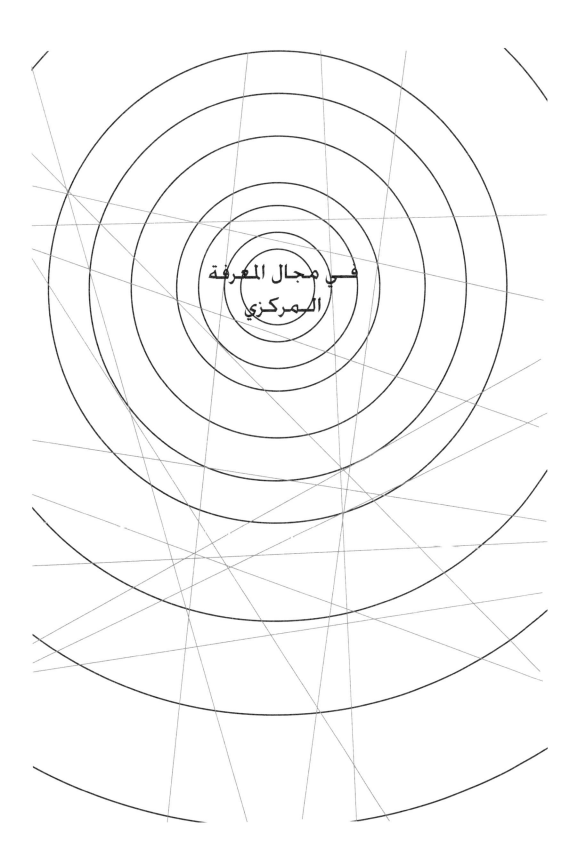

He for whom Allāh has not appointed any light has no light.[1]
[al-Nūr, XXIV: 40]

What would you say of one who seeks to divide light into discrete parcels and package it, in order to facilitate its consumption?

What would you say of one who enters into light, but comes out of it denying its existence, because he could not hold it or touch it or put it into his pocket?

What is the benefit for one entering into circles of light to partake of them, then, seeking increase, he continues on the same path by which he entered, and goes back out into the darkness, and says: "Where did that light go?"

He did not benefit from the light of the circle he was already in to set straight his path, and passed into the following circle.[2]

He will say, "My Lord! Why hast Thou raised me blind, when I used to see?" He will say, "Thus it is. Our signs came unto you, but you forgot them. Even so, this Day shall you be forgotten!"
[Ṭāhā, XX: 125-126]

1. [Editor's footnote]: All translated passages of the Quran are based on *The Study Quran: A New Translation and Commentary* [Seyyed Hossein Nasr, Caner Dagli, Maria Dakake, Joseph Lumbard, Mohammed Rustom, Harper Collins, 2015]. Some changes in the translation have been made in order to better reflect the author's preferences and the contents of this book.

2. [Editor's footnote]: As in the Quran, this book uses the pronoun "he" to refer to both the male and the female. Moreover, Arabic uses the word *insān* to refer to any human being, man or woman. In this respect, the author uses the pronoun "he" and the term "man" to refer to any human being, and modern references to gender will not be addressed.

في مجال المعرفة المركزي

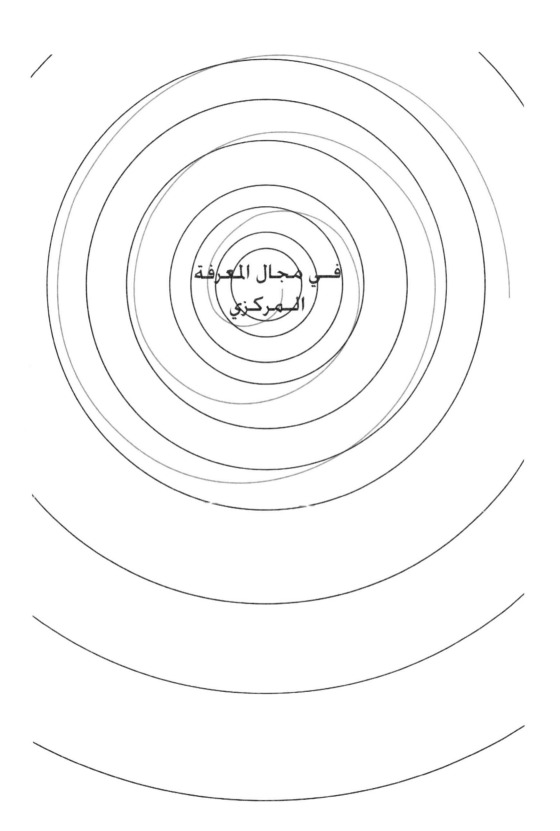

Author's Biography

Suleiman Sami Joukhadar (1956 - 2015) was born in Damascus into a family of scholars, including his great grandfather, Shaykh Muḥammad Joukhadar, one of the more recent scholars specializing in the science of Hadith (sayings of the Prophet, peace be upon him) in the Levant. His great grandfather was versed in the Hadith transmitted by Bukharī through a credible chain of narrators (*sanad*) and was also the Judge of the Levant.

His grandfather, Shaykh Suleiman Joukhadar, was an eminent scholar of his times; he was appointed a judge in the two holy cities of Islam, Mecca and Medina, and in the Levant. He founded the Institute of Legal Studies at the University of Damascus and also served as Minister of Justice.

His father, Dr. Iḥsānallāh Joukhadar, earned his International Doctoral Degree (Ph.D.) in Law from the Sorbonne University, France.

Through transmission from his family, Suleiman Joukhadar gained an in depth mastery of the sacred sciences such as the science of *Abjad* and numerology.[3] He authored many research papers on Islamic studies and sciences, in which he presented a forward-looking approach to Islam and to the Quran that reflected a long-term vision.

He conducted research within the field of historical architecture and plastic arts, which he studied at the Sorbonne. In this respect, he organized many exhibitions in which he presented his art work, expressing his thought and portraying his futuristic vision.

He wrote many articles in the fields of design, innovation, arts and sacred architecture. He masterfully expressed his thought in French, English and Arabic.

3. See definition on p. 137.

"...May the Mercy of Allāh be upon us and Moses; had he (Moses) shown patience he would have seen wonders..."

[Ṣaḥīḥ Muslim: 4386, the Book of Merits][4]

4. [Editor's footnote]: The numbering of the hadiths mentioned in this book refers to the Saudi Harf edition.

Preamble

A few years ago, I was asked to contribute to the first issue of an American academic journal specifically concerned with documenting and studying the religious, spiritual and metaphysical aspects of human heritage. The objective of this journal is part of a contemporary intellectual trend to which many associations and prestigious universities contribute, because they are conscious of the importance of documenting what has remained of customs and living oral traditions, before they become extinct through the invasion of modern civilization. I therefore contributed a sample article, in which I presented my own heritage. It was a unique contribution for our time in general, especially because it concerned the Islamic heritage.

When I wrote that mere two-page text, my sole concern was to document, in a succinct way, some very rare knowledge in order to preserve it from loss, with the intention of safeguarding it for anyone willing to decode its symbols and understand its secrets; as it was a very sensitive subject, I had to seek some advice. I then realized that many of my basic assumptions were not taken for granted by others, and that what was clear and lucid to my understanding was unclear and ambiguous or even completely unknown to others. Therefore, I had to write about these unknown details to complement the text; but I ended up rewriting the article, adding a few paragraphs to the original text and an introduction specific to the text itself, as well as another general introduction, addressing in particular the sacred sciences discussed in the text. **The text also stressed the importance of the Islamic heritage, which has often simply been neglected.**

It was also necessary to provide a preamble to the present treatise, the purpose of which is to establish mutual understanding with the reader by discussing points such as:

1. An overview of the structure of this work and the methodology used.
2. A discussion of some cultural and educational concepts and information that might differ from those of the reader.
3. The question of whether written sources define the Islamic heritage.
4. Challenges of conveying knowledge or thought in words.
5. A prelude to the general introduction to sacred knowledge, which constitutes the general framework of the subject matter of the present treatise.

1. An Overview of the Structure and Methodology Used

After the preamble, which is written as a sort of friendly conversation in modern language and tone, the reader will be treated to an introduction to sacred knowledge, in which the language and style will change, along with the tone and pace, to make it closer to the style of Sufi texts than to the Western and modern style. The reader will notice different levels in the information presented: some of it is intended for the select few, while other information is intended for people who are well versed in the Quran.

The main hurdle in the general introduction is in presenting certain ideas, which are in fact only logical introductory statements **to conclusions that are not explicitly presented**, but left unspoken, in a long presentation with elaborate explanations. These parts require the reader to remain alert and have the capacity to connect different pieces of knowledge together, in order **to be able to infer the meaning.**

The third part, in which the style and the language change according to the subjects presented, is an introduction to the *treatise* itself. The reader will be presented with necessary definitions, the methodology for approaching any treatise, and how to approach this treatise in particular. As an example of the changes in style, the two definitions presented at the beginning of the Introduction to the Treatise[5] are not intended for readers who are not acquainted with the subject matter of our book. They do not define or introduce any knowledge that the novice can take as a starting point for grasping the subject matter in question. In this case, for whom are these definitions intended? What is the point of presenting them? What does this usage imply? What are the essential allusions presented? And what are they telling us? On the other hand, there are (in the Introduction of the Treatise) clear definitions of *treatise* and *allusion*. The Introduction to the Treatise, however, is a simple and clear text that is not hard to understand. Reading it before the general introduction may help elucidate any ambiguities that may arise in the treatise itself.

The preamble is indispensable for giving the reader a proper approach to the whole book in general and to the two introductions in particular which are indispensable for the subject matter of the book, the treatise. If the reader finds it difficult to understand parts of the two introductions, he can go back to the preamble for answers. When sections of the treatise are not clear, answers will be found in the two introductions.

As I will mention later, the two introductions document some aspects of traditional oral teaching. They invite us to reconsider our way of thinking, correct it and develop it as a means of preparing the mind to benefit from sacred knowledge.

The preamble makes up one unit with the two introductions. This unit helps give the reader a profound comprehension of the subject matter of the treatise, or of any similar source. I therefore recommend reading the preamble and the two introductions carefully, a few times, and thinking deeply about their content **until**

5. [Editor's footnote]: see page 77 ff.

they become part of one's own knowledge and possible to make use of when reading the treatise itself.

Anyone who does not have good knowledge of the subjects presented in the treatise and is not well acquainted with them will find the major part of the treatise difficult to read and to understand in detail. The treatise is primarily meant as a means of preserving a heritage from loss and decay. It was not written with the intention of making the subject matter known or spreading its content. However, it does include sections that are clear and very important.

I am currently thinking of writing other books that could be added to the present one, so that they would all make one integral unit, providing the reader with the necessary material to enable him to understand deeply a vast aspect of knowledge that is close to becoming extinct. The style of these books shall be close to the style in which the present treatise is written, in which some sections are explicitly presented and defined and the others are platforms for oral teaching.

2. Review of Some Cultural and Educational Concepts

Some Western readers might be surprised that some of the knowledge they consider an integral part of their culture is actually presented in this book as part of the Islamic heritage and culture. The reason for this is that Western readers have been accustomed to viewing this knowledge from a certain perspective and conceiving of it in a particular way, to the extent that they instinctively refuse to consider it from other points of view. The problem becomes more serious in the absence of written documentation to prove that certain knowledge, which they consider Western, is in fact Islamic.

The answer to that problem resides in going back to a proper definition of the Islamic heritage and in meticulously checking the sources of information. If we read the Quran and the Hadith accounts in Arabic carefully and intelligently, we can easily prove that Islamic culture did not start with an individual and at an uncivilized point in time; and it was never a denial of the authentic, truthful human culture that preceded it. Islamic culture was never a copy of other civilizations nor a mixture of their legacies.

I am amazed to see how many people compare the first military campaigns by the Muslims to 'open up' new lands to the invasions that the falling Roman Empire endured at the hands of the Goths, the Visigoths, the Huns, the Vandals, the Germans and the Franks. The fact that these uncivilized tribes destroyed European civilization does not mean that the first Muslims did the same, merely because some of them were of tribal origins. The invasion of the Goths and the northern European tribes was barbaric, uncivilized and devastating and threw Europe into darkness and backwardness for long centuries.

On the other hand, Muslims turned Spain, which was backward under the Visigoths' rule, into a shining center of civilization. In less than one century, and after the first campaigns to open up new lands to Islam, Muslim civilization yielded tangible fruits and surpassed all the contemporary civilizations of the time. Could that have been possible if they had been simple uncivilized invaders?[6] Could that have

6. One of the unjust accusations leveled against Muslims, who deserve all the credit for transmitting Greek heritage, corrected and developed, to Europe, was that the Muslim 'invaders' were the ones responsible for burning the Library of Alexandria. It is important to note that none of the Christian historians of that time mentioned a word on the subject before it was mentioned for the first time five centuries after the conquest of Alexandria. In 933, Eutychius, the Melkite Patriarch of Alexandria, wrote a lengthy account describing the conquest of Alexandria and he did not mention the incident. Sadly, this lie was first mentioned by Ibn al-Labbād ʿAbd al-Laṭīf al-Baghdādī (1161-1231). I wonder who could possibly have whispered this lie into his ear when he visited Alexandria five and a half centuries after the conquest of the city. In any case, another historian took pains to complete the story with intriguing and touching details (like claiming that ʿAmr ibn al-ʿĀṣ gave his orders for the books to be used to heat Alexandria's four thousand bathhouses for six months!) but the orders were given by none other than the Jewish Syrian Bar Herbraeus, who converted to Christianity and was known by the name Abū l-Faraj (1226-1286). After using this lie to draw sympathy and nostalgia for the treasures of ancient human civilization, which were lost in this library, another lie emerged. This time they claimed that 'Amr ibn al-ʿĀṣ wrote to ʿUmar ibn al-Khaṭṭāb asking what he should do with the library and that 'Umar's answer was: if the books were not in line with the doctrines and teachings

been possible if they had copied the Torah and previous sacred texts? Could a copy surpass the original? If they were really uncivilized invaders, how do we explain their love for knowledge?

Let us stop for a moment and think of the mentality of these Muslims who opened up the world. Their reference was the Quran and the Hadith accounts.[7] What could be wrong with a mentality that is ordained to seek knowledge, as in this verse: ... *Say, "Are those who know and those who do not know equal? Only possessors of intellect reflect."*... [al-Zumar, XXXIX: 9], and this tradition, according to which, when asked about knowledge, the Prophet ﷺ (blessings and peace of Allāh be upon him)[8] answered: *"The superiority of the learned over the devout worshipper is like my superiority over the most inferior amongst you (in good deeds)."* [Sunan al-Tirmidhī: 2609]. These words could only have increased the interest and fervor of Muslims of that age for seeking knowledge.

What some find hard to understand is the outstanding result of this thirst and drive for knowledge, especially since these Muslim seekers possessed universal foundations and keys to sacred knowledge, which they received through revelation. What they achieved was so incredibly amazing that some scholars proposed theories without foundation just to prove the contrary, such as Dosso in his study of the Umayyad Mosque of Damascus, the symbol that embodies this leap of civilization.

The crusades were an incredible shock for the West. The crusaders, who returned to Europe after failing in their mission, felt a dire need to catch up with the Islamic civilization. Interest grew significantly in Islamic books and in translating them, by monks of the Benedictine Order in particular, with the help of the Jews of Spain. Needless to say, this interest and translation work must have saved part of the

---

of the Quran, they should be burnt and that if they were in line, then what benefit could there be in keeping them; they should be burnt! It is noteworthy that the first burning that caused the loss of the inestimable heritage of ancient civilizations happened upon Julius Caesar's invasion of Alexandria in 47 BC. The library was brought back to life again, only to be destroyed one more time in the wars between *Zenobia* of Palmyra and the Roman Emperor Aurelian and again by Emperor Diocletian. Indeed, the person responsible for destroying and sacking the library was the Byzantine Emperor Theodosius I in 391 as part of a larger effort to destroy all that was not in line with the doctrine of the Church. This unfair fanaticism was not an exception, but rather the rule: the Byzantine Emperor Justinian was the one who closed the famous Platonic school in Athens, and the schools of Alexandria for non-conformity with the Church. For the same reasons, the Byzantine Emperor Zeno closed the school of Edessa in 489 which had been famous for promoting Nestorian teachings and Greek culture throughout the East. And under his rule, all that remained of the books of Alexandria was destroyed. In 728, Leo III the Isaurian burnt thousands of manuscripts in Constantinople. These acts continued for long centuries under the Inquisition, following in the footsteps of Saint Paul, who ordered his followers in 54 AD in the city of Ephesus to destroy all books that contained 'heresies.'

7. To understand the mentality of these Muslims, the reader needs to know their beliefs. An index of the themes of the Quran alone would suffice to give an idea about the rich culture of anyone who reads the Quran. It is also important to note that these 'conquerors' had close exchanges with the most prestigious part of the world at that time, in the annual winter and summer trade journeys (cf. Quraysh, CVI: 2).

8. [Editor's footnote]: The expression 'peace be upon him' or 'blessings and peace of Allāh be upon him' translates the Arabic phrase: *ṣallā Allāhu ʿalayhi wa-sallam*, which is written in the form: ﷺ

Islamic libraries, which were destroyed by the Tatar and Mongol invasions. Many of these books were hidden in European monasteries and were considered dangerous. They were only made available to a handful of trusted clergymen.

Sadly, Islam and Muslims were considered the enemy and were thus hated. It was not safe to cite the enemy in Europe at the time of the Inquisition. It was safer, in the time of the Pope's 'Black Index' of forbidden books (Latin: *Index Librorum Prohibitorum*) to translate books or rewrite them, to alter the content and not mention the original version at all. All these precautions were not enough to protect Galileo from being accused of heresy for declaring that the earth revolves around the sun, and in 1633 he was forced to deny this belief on his knees in front of the Inquisition to avoid being burnt to death. The poor man remained banished from God's mercy and that of the Church until Pope John Paul II spared him at the end of the 20th century.

It is now widely known that Leonardo da Pisa, also known as Fibonacci, was the one who introduced the Arabic numbers to Europe as well as the famous sequence which bears his name. However, we do not generally know that this sequence, which is widely used in Ayubid architecture, for example, is originally Islamic.[9]

Harvey did not mention Ibn al-Nafīs al-Dimashqī and attributed the latter's discovery[10] to himself. Pascal did not neglect to express his hatred for Islam and Muslims,[11] but forgot to mention that the famous triangle he named after himself was in fact al-Karkhī's. Al-Karkhī's manuscript can be found in the National Library in Paris. The works of al-Ṭūsī appeared amazingly in Descartes' work centuries after his death. The same is true for al-Bīrūnī's works, which appeared in Newton's work. The astronomical discoveries of Abū al-Wafā' al-Būzjānī, related to irregularities in the motion of the moon, were attributed to Tycho Brahe, and the works of al-Battānī, like the law of cosines and tangents, were unjustly attributed to Regiomontanus, who attributed to himself the majority of Abū al-Wafā''s trigonometric theories and compiled them in his book *De Triangulis*. The invention of the pendulum was attributed to Galileo while the truth is that Ibn Yūnus al-Sadafī had invented it six centuries earlier and was the one who first proposed the idea of logarithms, which was attributed to John Napier. Nor was Copernicus embarrassed to claim Ibn

9. See the definition of the Golden Ratio on p. 138.

10. [Editor's footnote]: The discovery of the right-sided pulmonary circulation.

11. In Pascal's most famous book *Les pensées*, which is considered one of the principal French and Western intellectual classics, he draws a comparison between Jesus and Muḥammad ﷺ in paragraph 599. After he presents Muḥammad ﷺ as a murderer, he claims that Muḥammad ﷺ banned reading, whereas the apostles were ordered to read. Pascal's reference was none other than his fellow countryman Montaigne, who declares: "According to what I have heard, Muḥammad used to ban his men from acquiring knowledge." Maybe Montaigne's source was none other than his own mother, whose Jewish family escaped Spain. The first word of the Quran revealed to Muḥammad was: "Read" while – not long before Pascal was born – the punishment reserved for anyone who dared to read any translation of the Holy Bible other than the official Latin translation was being burned to death and eternal punishment in hell. It is also important to note that the few who could read Latin were either from the clergy or the highest class of society. Also, the volumes included in the 'Index,' the black list of banned books, are stunning.

al-Shāṭir al-Dimashqī's great discoveries for himself. Ibn al-Shāṭir's manuscripts can still be found in Poland, Copernicus' homeland.

Any serious, contemporary researcher knows that these examples are only famous cases in a long list, which does not include books by Muslim authors that are still hidden and surviving texts that have been distorted. For example, the Tarot is despised by some and respected by others. Many scholars have carried out extensive research and studies on the Tarot,[12] but they rarely mention that it was the Arabs who introduced it into Europe in the 13th century, as Gérard van Rijnberk confirmed. It is certain, however, that the Arabs did not introduce the Tarot to Europe with the distortions that it underwent later on. Why should it be strange, therefore, that up to this day, Muslims hold the correct keys to decoding the symbols of the Tarot? Their heritage is universal and pure and enables them to understand accurately the true message of the Tarot and also of the Tetraktys,[13] for instance.

Again, it seems as if the Tetraktys has nothing to do with the Islamic heritage. I personally know Westerners who claim that they received an oral tradition going all the way back to Pythagoras and the Platonists. They consider themselves the legitimate heirs of the Greeks in the sciences of the elite. They would be so embarrassed to see a Muslim present a reading of the Tetraktys from the point of view of the Islamic heritage. Here again we face the problem of unwritten references. However, a simple review of some neglected facts helps clarify things.

Civilization did not start with the Greeks. The prototype of Greek temples was found in Emar, the Syrian city that flourished on the Euphrates in the second millennium BC. The caisson, which is considered a unique detail in Greco-Roman architecture, was first used, according to modern archeological findings, in the second millennium BC in the city of Mary on the Euphrates, which is located in Syria today. It is peculiar that art historians called some of the Egyptian columns by the name 'Proto-Doric' or 'Pre-Doric' columns. These columns first appeared in Egypt when the peoples of Greece were still pre-historic. Why do we not call the Doric columns 'post-Egyptian columns'? Of course one would not approve of that term.

It is well known that Cadmus the Phoenician, or more accurately the Canaanite, was the one who founded the city of Thebes and taught the Greeks the alphabet that existed on the Syrian coast at the end of the fourth millennium and the beginning of the third millennium BC. Europe was named after his sister.

Pythagoras was the son of a wealthy merchant who traveled extensively in Sidon, according to his oldest biographies. Sidon, the Phoenician city, or rather the city of the Canaanites, was one of the most famous cities on the Syrian coast. In Samos, Pythagoras' first teacher was Pherecydes of Syros. Thales advised the young Pythagoras to go to Egypt. He first went to Sidon on the Syrian coast, before moving to Byblos and Tyre. At the time, both were famous cities on the Syrian

---

12. See the definition of the Tarot on page 135.
13. See the definition of the Tetraktys on page 134.

coast. He then left for Egypt where he settled for twenty two years before being captured and taken to Babylon by the soldiers of Cambyses. After twelve years, at the age of fifty six, he returned to Samos to teach what he had learnt from the Near Eastern civilizations. He had a tremendous influence on the Greeks and his ideas had an important place in the works of Plato and the new Platonic school or the School of Alexandria:

- Porphyry said: "In his fifth book of stories, Neanthes wrote that Pythagoras was a Syrian citizen of Tyre."
- Pythagoras did not invent the seven musical notes; they existed long before him in Mesopotamia, as has been proven by archeological findings.
- Pythagoras did not invent the triangle that was named after him; it was known long before. It would be more accurate to call it the Egyptian triangle.
- Did Pythagoras invent the Tetraktys? Definitely not.

The Tetraktys was carved in a slightly different way in the majority of the sacred buildings in Syria in particular, like in examples from the second millennium BC up until the time of the Romans under the rule of the Syrian dynasty, and in Palmyrene architecture. This architectural detail reappeared again under the Islamic empire, the Umayyad caliphate, of which Damascus was the capital, and continued to be used in Islamic architecture. In any case, the center of Pythagorean and neo-Platonic thinking flourished in the region between Alexandria and Antioch. The most brilliant representatives of this school of thought were famous names like Numenius of Apamea (central Syria), Nichomachus of Gerasa (south of Syria), Iamblichus of the Bekaa Valley (west of Syria) and Antonius of Latakia (Syrian coast).

If it is acceptable that a British person, a European or a Goth claims to be the heir of the Pythagorean heritage[14] why is it not acceptable for a Muslim from the Near East to present a reading of the Tetraktys that uses the universal and accurate keys provided by the Islamic heritage? This is especially true since the torch of knowledge and thought was soon held by Muslims in the first decades of Islam at the time of the Umayyads. They dutifully carried it for many centuries, during which, after having stagnated, human thought made giant unprecedented leaps.

The truth is that the main principle in Islam is Unity. Muslims know that the Truth is One. Truth manifests itself in many ways and on many levels. To Muslims, Truth is manifested in many ways in the sources of revelation, which they have preserved

14. As an example of this, let us look at the oath sworn by the members of a research and study group of European Civilization (Groupe de Recherche et d'Etudes sur la Civilisation Européenne: GRECE). This is a school of thought founded by the French scholar Alain de Benoist. The members of this school, the majority of whom were French, swore this oath during a pilgrimage to the city of Delphi in 1972: "We are Hellenists, Italians, Belgians and French gathered today under the banner and the care of the 'god' Apollo with all our brothers the Europeans. We swear to work with all our energy and all our will to revive European culture. In this sacred place in Delphi, long a symbol of our world, we swear allegiance to our heritage and allegiance to our children." The French historian Odon Valet commented on this oath saying: "This sentimental approach legitimizes the appropriation of a civilizational heritage exclusively by Europeans, on the condition that we forget that the thieves who sacked the sacred treasure of Delphi were the Gauls (the native inhabitants of France)."

in purity and without any alteration. They can use *qiyās* (analogy) to understand any manifestation of the Truth in a profound, accurate and detailed way. An accurate and intelligent reading of the Quran and the Hadith in the Arabic language proves that the Islamic heritage is a continuation, purification, and completion of the entire universal heritage that preceded it:

- Muslims ask the Lord at least 17 times every day in the five prayers while reciting the first surah of the Quran: *Guide us upon the straight path, the path of those whom Thou hast blessed...* meaning the peoples who were before us and followed the path of Truth.
- The first of the long surahs of the Quran begins: (1) *Alif Lām Mīm* (2) *This is the Book in which there is no doubt, a guidance for the reverent,* (3) *who believe in the Unseen and perform the prayer and spend from that which We have provided them,* (4) *and who believe in what was sent down unto thee, and what was sent down before thee, and who are certain of the Hereafter.* [al-Baqarah, II: 1–4]

This is further evidence of Muslims' attachment to the knowledge of the peoples who received tidings of the Truth and that Islam is a continuation of the path of the people of Truth.

In the Arabic language, a Muslim cannot utter the name of a prophet without mentioning the words 'our Master' (*Sayyidunā*) before it and the expression 'peace be upon him' after it. Both phrases are used together to show the great reverence Muslims feel towards all the prophets, and the best example of this is the following verse from the second surah of the Quran:

> *The Messenger believes in what was sent down to him from his Lord, as do the believers. Each believes in Allāh, His angels, His Books, and His messengers. "We make no distinction between any of His messengers." And they say, "We hear and obey. Thy forgiveness, our Lord! And unto Thee is the journey's end."* [al-Baqarah, II: 285]

Another confirmation of the same idea is found in the third surah:

> *Say, "We believe in Allāh and what has been sent down upon us, and in what was sent down upon Abraham, Ishmael, Isaac, Jacob, and the Tribes, and in what Moses, Jesus, and the prophets were given from their Lord. We make no distinction among any of them, and unto Him we submit."* [Āl 'Imrān, III: 84]

The Prophet ﷺ insisted on this when he asked his companions not to consider him better than Jonah, as in the Hadith: "*And I do not say that there is anybody who is better than Yūnus (Jonah).*" [Ṣaḥīḥ al-Bukhārī; 3162]. In fact, Muslims venerate all prophets without any preference. The prophets are the people of faith and of true knowledge. They are one nation. In the 21st surah, some of them are mentioned together with the main events in their lives. Then Allāh concludes by saying: *Truly this community of yours is one community, and I am your Lord. So worship Me!* [al-Anbiyā', XXI: 92]

3. Do the Written Sources Define the Islamic Heritage?

The main subject of this book is knowledge that I was being prepared to receive from my early childhood. At that time I was given in a very subtle way, the essential keys to some of the sciences of the elite. At the age of 14, I made my first steps into that knowledge and, for my spiritual journey, I was initiated in a direct manner, and in ways that were very particular.

Time passed and I had accumulated some rare knowledge that I shared with university professors in the United Kingdom, France, Germany, the United States, Egypt, Lebanon and Syria, as well as some religious and intellectual figures. That was in 1980.

A few years ago, for a variety of reasons, I felt the necessity to publish part of the knowledge I had received orally. For example, when I read studies that were based on the Islamic heritage or related to it, or met experts working on these themes, I was shocked by the way these themes were approached and by the conclusions drawn. This was particularly the case for studies related to the sciences of the elite, spiritual philosophy and even fields like Islamic architecture.

I was often amazed at the glaring ignorance of some specialists in studies and research related to Islam and Islamic civilization. How is it possible that such experts claim to have enough authority to take on the responsibility of teaching subjects related to Islam at the university level when they cannot read a single sentence in Arabic? How is it possible that they make the claims they do when all they rely on are translations and they are not capable of using original sources? Many of those who are considered reliable authorities in subjects related to Islam pass judgments while they are unaware that their judgments are their own speculations, not the truth.

I once met a famous university professor, who is considered an authority on Islamic art and architecture. He also teaches other subjects and supervises research at the PhD level on topics that are very often related to Islamic thought and philosophy. He usually cites Ibn al-'Arabī and it impresses his entourage. I was surprised and disappointed to know that he was completely ignorant of many of the fundamentals that are essential for understanding any knowledge related to Islamic spiritual philosophy or even to Islamic art or sacred architecture. He cannot read one sentence of the original Arabic text of the *Futūḥāt al-Makkīya*. Nevertheless, he wrote one commentary after another on Ibn al-'Arabī's *Fuṣūṣ al-Ḥikam*. If he were an amateur or had been teaching in any other field, he could have been excused for this.

It is really disappointing that many Muslims have been influenced by the opinions of these 'experts' and forged their own convictions under their supervision in Western universities. The problem is that the sources available to those Western researchers are translations of Arabic books that represent only one aspect of the Islamic heritage and culture. In addition to these translations, there are so many studies on Islam of varying quality; some are wonderful, while others ... defy description.

When faced with this flood of studies, the only way that the Westerner has to judge is to compare them to his own mentality and culture, which are deeply influenced by the Holy Scripture and the Church and by what he was taught and what he knows of other civilizations. The main problem that such a researcher faces is the authenticity of the Islamic sources and references. In this case, one can resort to the very scarce oral sources that are very difficult to find and the written sources, which only cover some aspects of the oral sources. Because of the dramatic events inflicted upon the Islamic world, it is nearly impossible to recover the entire, authentic Islamic heritage.

I am a Muslim from the city of Damascus, the capital of the Umayyads, a name closely linked to great names, such as the Prophets Abraham and Elias, Elisha, John, and Jesus, peace be upon them, as well as Apollodorus the Damascene and many of the companions of the Prophet Muḥammad ﷺ, like Bilāl and Abū al-Dardā', may Allāh be pleased with them. In addition to these, Damascus is associated with many Muslim scholars and people of knowledge (*'ārifīn*) like al-Ghazālī, Ibn al-'Arabī, Rūmī, Shams Tabrīzī, Ṭūsī, Ibn al-Shāṭir, Ibn al-Nafīs, al-Farābī, Ibn Kathīr, Ibn Taymiyya, al-Jaldakī and others.

The Islamic culture that the best of the Islamic Damascene intellectuals possess nowadays rarely goes beyond printed Arabic books. This is a wonderful thing, but this does not mean that what they ignore or what does not exist in print never existed in the Islamic heritage. I started noticing this fact when I realized that the Islamic heritage related to sacred architecture was completely lost after World War II in my city. The heritage related to the science of letters is also being lost, and especially the oral heritage related to the science of numbers.

I am well placed to claim this, thanks to the knowledge I received from my family. My grandfather was one of the highest authorities in Islamic affairs in the Ottoman Empire, as was his father, Shaykh Muḥammad, who was a judge in the Levant, a shaykh and an authority in the science of Hadith and who withdrew into retreat in the Umayyad Mosque. He was given the title of Abū Ḥanīfa Junior, and was a disciple of Ibn 'Ābidīn, the author of *al-Ḥāshiya*.

Members of my family also received sacred sciences of the elite, given that they are descendants of al-Ḥasan through Shaykh 'Abd al-Qādir al-Jīlānī and the chain of transmission of Ma'rūf al-Karkhī, Ḥasan al-Baṣrī and al-Junayd. Sadly, I am the only person in my family who possesses what was left of this heritage. Many of those who received sciences of the elite did not have a chance to transmit their knowledge outside of their restricted circles, as I have been able to do. I met only a few people, who are completely unknown, who possess what has remained of some of these precious sciences.

Lack of sources does not mean that a heritage pertaining to a specific field does not exist. I think it is naïve to imagine that the published works of famous authors like al-Rāzī and Ibn Sīnā are in fact the real core of the sources written in the Islamic civilization. The fame of some authors is not a guarantee that their famous books are the best that exist in a specific domain. Many great knowledgeable scholars were not known for their real worth and were completely unknown to fellow scholars of their age, because they dedicated their work completely to Allāh and did not seek privileges from people or rulers.

I was told that a disciple of Shaykh Badr al-Dīn al-Ḥasanī went to see him and found his master crying. The shaykh told his disciple that his days in life were numbered and that he had seventeen sciences that nobody had asked him about; not even one question. He was afraid that Allāh would take him to task for this responsibility. Before his death, because he did not have the permission to unveil that knowledge, Shaykh Badr al-Dīn ordered all of his books to be burnt, as did so many great scholars. Despite his vast knowledge in the sciences of the elite, my grandfather did not record any of it in writing, whereas his friend, Shaykh Aḥmad al-Ḥārūn, wrote many books that are in the possession of one of his heirs. This heir still does not dare to use them in any way. These three men passed away only a few decades ago in Damascus and are totally unknown. What about the other unknown scholars in the Islamic world?

In fact, the books written within the Islamic heritage can be divided into two categories: those which were meant for a general readership and others that were transmitted through an exclusive chain of transmission. The authors of the second kind of books were only concerned with preserving the knowledge they had received. Therefore, they took some precautions when writing, even though their books were not intended for the general readership. They wrote in a way that can only be accessible to those readers who were orally given the keys to unlock that knowledge. These keys would decode that knowledge, if Allāh wills. Generally, there would only be one copy of a book on a given subject. As a general rule, the written sources were composed for specific mentalities in specific countries and times, which influenced their content, form, style and even the wording of the texts.

Does knowledge of Arabic or Persian suffice to approach these books and understand them? According to their original intention, reading these books was a task assigned to chosen disciples in the presence of a shaykh. This supervision allowed the shaykh to help his disciples acquire only what should be acquired from the book and reformulate it to give them the correct understanding to complement what they already knew within the framework of the general topic as they received it. People who had the chance to have access to knowledge in this way can see the difference that exists between the exact meaning of what they learned and a study that was made of the same subject, but done by an 'outsider.'

We live in an era in which we are confronted with a deluge of printed material. This begins in school, where the intelligent student is given the same treatment as the simple-minded one, in order to give both a fair chance. The student is used to studying books that are imposed upon him and on which his future depends. They are so simple, so clear, that they suffocate the student's own capacity to draw conclusions; books that face the student's wrath if they do not bow to him! That is because the textbooks are supposed to have all that he needs. The student does not learn to get up from his seat to find the missing pieces in the books. As his personality takes shape, the first things he reads most of the time are entertaining literature, such as newspapers, magazines or novels. He is soon accustomed to a fast rhythm in reading and is only concerned with following the flow of the events and knowing what happens on the next page. This style in particular is the habit of most people in our age: fast, dispersed, shallow reading, justified by lack of time.

We live in an age when a deluge of words is used by publishers, the press and in particular by media outlets to express very simple ideas. This has made people's minds thick and has blurred their sense of understanding and weakened their spontaneous and dynamic thinking in the face of a substantial text with words and concepts that require pauses for deep thinking.

Is the reading capacity of a person from our time the same as that of a reader of one thousand years ago, who read a precious manuscript in peace, repose and with an abundance of time? The way scholarly texts are read nowadays is something beyond true seekers' imagination. The correct reading of a sacred text cannot be carried out without a long and meticulous preparation of the disciple. The disciple is taught how to elevate himself to the level of the knowledge he is hoping to receive. The disciple is taught that his understanding changes according to his state. This is because his effort is useless if he understands the words, but not the meaning behind the words. In order to quench his great thirst for understanding, the disciple can never prepare enough for this mission. He avoids anything that might blind his inner eye. He knows very well that it is not at all acceptable to seek knowledge through reading a book without putting into practice the proper *ādāb* (manners) he learned from his shaykh, such as being in a purified state of ablution and clean, in appropriate clothes and setting, after having prayed and invoked for a little while for success and realization, and more importantly after leaving behind all traces of his own ego.

The main difference between the knowledge and sciences of the elite and what is available to the general audience is not in the written information, but rather in the person receiving it. The difference lies in the skills developed, the lucidity and openness of mind of the seeker, and above all, in the inspiration he constantly seeks. What is the use of a library full of books in the absence of a light to illuminate its darkness? Knowledge without inspiration is more like a body without a soul; it soon perishes. Seekers never lose sight of the verse: *He for whom Allāh has not appointed any light has no light.* [al-Nūr, XXIV: 40]

Contemporary researchers have never been taught the technique of reading the books that contain the sciences of the elite. They do not possess the keys to unlock their mysteries. They do not distinguish that part of the text which is meant as a trap for intruders, from the sections which are part of a coded message. Neither can they differentiate between a technical passage that should be understood literally, and one that should be understood with an intuitive sense of the sacred, and a third that is intended for someone who masters the meaning of the symbol, how it is understood and how it is used.

What I have just mentioned applies to the case in which the researcher succeeds in working with a reliable source. However, is all that is written trustworthy? Do all the sources represent the heritage in the same way? Does what was written cover the oral heritage? Are the books that are still in existence all that was written?

I am the heir of what remained of a large private library that was burnt when the French protectorate army savagely attacked the old Damascene neighborhoods in 1925. This library contained manuscripts that survived other catastrophes. What can we say about thousands of other manuscripts that perished because of negligence or in the incidents or invasions that swept over the Islamic world, or those that were destroyed by the Inquisition or by other forces? The written sources that are still in existence today are only a small portion of the number of books burnt.

The books published in our time are an insignificant fraction of what survived. The majority of the unpublished manuscripts are scattered in famous libraries, especially in private, or very private collections, to which access is restricted to a very limited number of individuals. This is happening outside of any kind of supervision or approval of Islamic authorities.

I know many sincere researchers who understand the challenges presented by the written sources and that the solution is to resort to oral sources, so they have tried to get in touch with some knowledgeable people in the field. Unfortunately, these researchers were not prepared for the encounter with a knowledgeable shaykh or aware of how to recognize a real shaykh among many false ones or how to behave in his presence. When I talked to them, it was clear to me that they did not have a precise idea about the mentality, the abilities and the ways of the knowledgeable shaykhs. If they do not send away someone they do not want to meet, then these shaykhs receive the person for a particular purpose. Most of the time they hide their true nature behind some confusing appearances and contradictory answers. They do not answer the questions, they answer the questioner. They give every person the answer he deserves. Sometimes they even pretend not to know anything. It is hard to penetrate their mysterious world, but they do not hesitate to provide an opportunity, or a piece of advice or an allusion, to the person they meet. But the latter rarely understands the allusion and rarely takes advantage of it.

The truth is that there is a lacuna or a great lack in the written sources, which can only be compensated for by the oral sources.

Our life style and all the novelties that have penetrated it and affected it have rendered traditional teaching an almost impossible process. As a result, the scarce depositories of this heritage pass away without leaving successors to carry on the torch after them. At the same time, interest in subjects related to the Islamic heritage has increased on the part of modern universities and the authorities linked to these universities. The main problem therefore, as I mentioned above, resides in the sources. This is one of the reasons that made me feel the necessity to try to put part of what I inherited into writing.

4. The Challenges of Communicating Knowledge or Thought in Words

As a first attempt, I thought of writing a treatise about a very vast subject, the **3×3 Wafaq square[15] and its relationship to prayer**, one of the five pillars of Islam. I chose the subject because of its importance and its comprehensiveness and in order to record that which had not previously been written.

However, at this very moment, I am overcome by a feeling of confusion as I try to translate part of 'an indivisible whole' into a sequence of discrete parts or crumbs. I am also trying to translate abstract, infinite ideas and concepts into expressions and words that are mundane and limited.

In fact, it would have been much easier for me to write about art or architecture or any other subject. This is because the language required for these subjects and their expression is common knowledge and should not be a source of confusion, especially since others have delved into these subjects and it would not be hard to find a model to follow and to take as a reference point for a common language of understanding. Art and architecture are widespread subjects that do not pose great linguistic challenges when approached within the framework of a common type of thinking. The problem arises when a person, who has received a kind of knowledge in a particular way, is asked to express it in another manner that is essentially different, especially in a confining means of expression such as that of writing. That challenge is made even greater when the content to be published is not a topic meant to be published, but rather to be transmitted from one individual to another, perhaps in a special way, as supported by Plato in his seventh letter, or in other more direct and elevated ways. In his seventh letter, Plato wrote:

> ...there is no way of putting it [this kind of knowledge] in words like other studies. Acquaintance with it must come rather after a long period of attendance on instruction in the subject itself and of close companionship [between master and disciple], when, suddenly, like a blaze kindled by a leaping spark, it is generated in the soul [of the disciple], and at once becomes [between master and disciple] self-sustaining.

Transmitting knowledge in a situation such as the one described by Plato has a great chance to be accurate and successful because the master carefully chooses a student who is capable of receiving the teachings in question and constantly prepares him for elevation to a level worthy of that knowledge, and gradually passes it on to him according to his individual state. All these ideal circumstances are not available in publishing, and the mission becomes more like sending a messenger to the dungeons of Babylon, rather than to its tower.

Given the purpose of this book and its content, and before I continue with this subject, I find myself compelled to subdue the style and the language of the book completely and put them at the service of the ideas and knowledge presented, and the philosophy related to them. It is my choice to place the style of writing at the service of the idea.

---

15. [Editor's footnote]: See the section on the Wafaq square in the Introduction to the Treatise, p. 77.

In the face of a language that is constantly changing, the reader's reference point should be the ideas, not the words. This is a very important point to highlight without stumbling into a style of language which constantly requires the reader to comprehend in an intelligent and flexible way.

This book should be approached in a comprehensive manner. **It is not at all appropriate to approach it in a systematic, consequential, critical, step by step reading.** The book requires multiple readings and many pauses to ask questions **without hastily seeking answers**. It demands pauses for thinking, meditating, linking information, inference and attempting to understand the words and information, unlike the way people are used to reading, but according to what I am trying to convey. The key to attaining this understanding is to:

- Break free from mental rigidity and existing prejudices
- Break free from the prevailing ways of thinking
- Rouse oneself and move within the book with flexibility, intelligence, and sincere intuition.

Otherwise, the book will remain strangely mysterious, and will not be understood, because the subject matter presented is knowledge that cannot be completely grasped without oral teaching.

In this context, writing is merely an attempt to preserve a heritage on the one hand, and on the other an invitation to go back to origins. Despite the fact that I studied at the Sorbonne and am acquainted with the academic style of writing, I have to say that I found it inappropriate to use this style in writing about a subject related to a heritage different from that of Western civilization. Because what I want to document is knowledge that is part of the heritage of learned Muslims, I am compelled to use their ways of communicating their ideas, out of devotion for and allegiance to their method and so as to deliver the message in the most appropriate manner. I am well acquainted with their styles, which are coherent within the purpose and philosophy of the subject at hand, but also pose some challenges. Furthermore, because the reader is not accustomed to them, the style might remain beyond his capacity of understanding.

Introducing and explaining this writing may provide an example of an accurate approach and a correct understanding for topics of this kind in any subject. The introductions and explanations may also be considered as a kind of translation and documentation of part of what is provided in the oral tradition of teaching, especially for beginners upon the path.

In fact, the use of storytelling, symbols and allusions is one of the subtle ways of communication used by the teachers of thought in the Near East. These masters are well aware that some ideas cannot be expressed if conveyed by merely an expression or a sentence or even a paragraph. These ideas can be expressed in an implicit way through many sentences or paragraphs that, if taken separately, would not have any meaning at all.

> **The main ideas transcend words, sentences and paragraphs that signify and suggest, but do not explicitly express the content in question.**

These masters know very well that words fail to express what they want to convey. Their message is a spiritual state and a lucid vision of Truth, not of ideas that are so small that they can be expressed by the words of daily life.

> **What is important in their eyes is the person who is receiving knowledge. What they care about is the intelligent person who can understand the allusion. That is because jewels cannot remain hidden in front of those who know their value. This is why masters do not shed light on the jewels and do not compel the common people to take them. Great masters are satisfied with signaling to those who are suited to such jewels of knowledge. This is the main characteristic of their style.**

5. Prelude to the General Introduction to Sacred Knowledge

Is knowledge limited to the general boundaries of the sciences of our time or does it go beyond them?

Beyond? The first thing that comes to mind is of course the fields of scientific, material, and experimental knowledge, because we are used to hearing about endless discoveries that stretch the limits and horizons of this knowledge.

Beyond? We speak of intellectual knowledge with less enthusiasm, because we have become used to novelties in opinions and theories and in points of view. We consider every novelty an addition to former attempts to reveal aspects of the truth.

Beyond? We look at the sphere of religion or canonical law with a lack of interest because we do not expect anything new, if it is not providing details, explanations, corrections, reminders, rephrasing, or repetition of what already exists. Anything else is considered unorthodox or heretical.

In these three instances, while seeking wider horizons, our thought never crosses the broad boundaries of what we already know. This result confines one within narrow boundaries, which become a trap in which one is pushed by one's ego to be content with what one already knows. One becomes the prisoner of one's own constraints. This leads to a hardness and seclusion of the mind.

For the individual to escape the trap of this vicious circle, he needs to have a large capacity to review, revisit his own knowledge and be able to push aside the veil that covers his inner eye. This reviewing needs to be done in the two directions in which the Truth manifests itself: outwardly (to the universe around us), and inwardly (into our own souls).

The problem with this reviewing is that the individual does not see what he does not know. This means that there has to be an outside source of knowledge that is *not* human, so he can learn how to see things from an angle different from his own. This is the reality of things; how precious is that outer source that we receive from the Creator of Existence Himself![16]

The knowledge received from Allāh contains the correct information in the three spheres of knowledge mentioned and even goes beyond these spheres, **as it includes all that can possibly be known, given that it carries the absolute foundations, laws and rules for all knowledge**; and that is the most valuable thing that a seeker of knowledge is looking for.

This divinely given knowledge is sacred knowledge, in other words, knowledge that is infallible, clear of any defects, distortions or errors. It is true knowledge that,

16. Needless to say, anyone who has free will and reason has the right to doubt things, because so many have told lies which they attributed to Allāh. However, when one meditates on the Revelation which is described as something that is "without a doubt" one would find the guarantee of truthfulness in the Revelation by its own words: *Alif Lām Mīm* and that a deep meditation leads to: *This is the Book in which there is no doubt....* [al-Baqarah, II: 1-2]

even if common people do not acquire it, the elite cannot live without it, because it is that which justifies and gives existence to man's vicegerency (*khilāfah*).

Is not religious knowledge, as part of the sciences of the Islamic canonical law, sacred knowledge? If we look more closely through the divine reference, *in which there is no doubt*, we will find that the limits of this knowledge go far beyond what religious teachers of all peoples teach.

We are certain that the Quran is a divine text whose verses are perfect, which means that it is not deficient in any way, nor does it include any unnecessary additions. Every detail that it contains is essential and indispensable. Based on that certainty, we have the right to ponder the reason behind hidden meanings and signs[17] in many of the details we are given and which surpass the domains of the *sharīʿah* and good advice.

Are differences to be found in the Quranic text? Are variances of importance to be found from one verse to another? Of course not! Are the details given in the story of Solomon in Surat al-Naml only used to further embellish the setting and make it more exciting, or do these details carry a message to the people of intellect? In that story, moving the throne from one place to another is something that goes against all physical rules. The creature who performed this act had knowledge of the Book. What could this knowledge be? What was this knowledge that David and Solomon (peace and blessings be upon them), possessed? What was that knowledge possessed by al-Khiḍr,[18] Moses, Jesus and all the prophets (peace and blessings be upon them) – beginning with Adam who was linked to knowledge, when he was first mentioned, and qualified as a vicegerent (*khalīfah*), until the last prophet, Muḥammad ﷺ? We find that all the prophets and messengers share a divine knowledge which surpasses the limits of canonical law. This knowledge was not exclusively given to prophets, but was passed on to some of their heirs. Some of these heirs preserved that knowledge while others abandoned it.

If we accept this as a fact and we are convinced of the necessity to acquire this knowledge that is linked to man's vicegerency (*khilāfah*), we need to ask about the means of acquiring it.

The best reference we can resort to, as mentioned before, is the Quran. If we search the Sacred Book, we will find the answers in a wonderful story, which is the only example of how to teach a chosen person sciences of sacred knowledge. The story of al-Khiḍr and Moses is the sole testament of this teaching in the Quran. Here we find a person of knowledge (Moses), who has a prearranged appointment with another person of knowledge (al-Khiḍr) searching for him to become his disciple and be taught correct guidance and knowledge received from the Presence of the Knower, the Wise.

It is interesting to note that, according to the conventional traditions of the readers, the story falls exactly in the middle of the Quran, which shows the importance of

17. [Editor's footnote]: "sign" translates the Arabic word *ishārah*. In other context, the word "allusion" is used.

18. [Editor's footnote]: Al-Khiḍr is a mysterious esoteric figure who guides Moses to knowledge, as the story is related in Surat al-Kahf.

the account. This part starts with the verse: *Did I not say unto thee that thou wouldst not be able to bear patiently with me?* [al-Kahf, XVIII: 75] Another hint of the importance of the story is that the number of times the word 'sea' (*baḥr*) is mentioned in the singular form in the Quran is 33. The middle three occurrences, numbers 16, 17 and 18, all fall in the story of Moses and al-Khiḍr and the middle occurrence of the 33, which is number 17, is followed by the word 'wondrous' (*'ajaban)*.

The importance of the story is highlighted by Hadith accounts in the two Ṣaḥīḥ collections, where we can find precious details in our search for sacred knowledge. All of these Hadith accounts start by relating the causes of the event, which were simply left out in the Quranic account:

- *Allāh's Messenger ﷺ said, "Moses got up to deliver a sermon before the Children of Israel and he was asked, "Who is the most learned person among the people?" Moses replied, "I am the most learned." Allāh then admonished Moses, for he did not ascribe all knowledge to Allāh; then came the Divine Inspiration: "One of Our slaves at the junction of the two seas is more learned than you."* [Ṣaḥīḥ al-Bukhārī: 119]
- *Moses had been delivering sermons to his people reminding them of the days of Allāh, His Mercy and His Wrath. And he made this remark: "I don't know any person upon the earth better than me or more knowledgeable than I."* [Ṣaḥīḥ Muslim: 4386]
- *While Moses was sitting in the company of some Israelites, a man came and asked him, "Do you know anyone who is more learned than you?" Moses replied, "No." So, Allāh sent the Divine Inspiration to Moses: "Yes, Our slave, al-Khiḍr (is more learned than you)."* [Ṣaḥīḥ al-Bukhārī: 72]

This brings us back to the question at the beginning of this chapter of what is 'beyond' in the sphere of knowledge. The Hadith teaches us a very clear lesson: it warns us of the danger of denying the existence of knowledge outside the rigid boundaries of what we know or far beyond them. In so doing, the individual cuts himself off from the path of good and confines himself within boundaries that he established. Little time passes before it becomes a prison for him and he becomes hardened and secluded therein.

What is most interesting in the following Hadith, is that the Prophet ﷺ wished for events, which had taken place one thousand eight hundred years before him, to have ended differently:

> *"May Allāh bless Moses, I wish Moses could have shown more patience so that a fuller account of their story could have been related to us."* [Ṣaḥīḥ Muslim: 4385]

In this unrealizable wish was a subtle hint from the Prophet ﷺ about the fact that what al-Khiḍr presented was unfinished and has a continuation. The allusion should be clear to anyone using his intellect. In another Hadith in the same chapter in Ṣaḥīḥ Muslim (the chapter on the merits of al-Khiḍr) the Prophet ﷺ said:

> *"May Allāh have mercy upon us and upon Moses. Had he shown patience he would have seen wonders, but fear of blame, with respect to his companion, seized him and he said: 'If I ask anything after this, keep not company with me. You will then have a valid excuse in my case,' and had he (Moses) shown patience he would have seen many wonders..."* [Ṣaḥīḥ Muslim: 4386]

What is striking is the repetition of the phrase "*he would have seen many wonders.*" What wonders would they have been?

> "*...had he shown patience he would have seen many wonders...*"

General Introduction to Sacred Knowledge

And indeed We have employed every kind of parable for mankind in this Quran. And man is the most contentious of beings. [al-Kahf, XVIII: 54]

"My son, if you want to read the Quran, read it as if it were revealed to you."
Muḥammad Nūr to his son, Muḥammad Iqbāl

The Story of Moses and al-Khiḍr (Peace and Blessings Be Upon Them)

In the Name of Allāh, the Lord of Mercy, the Giver of Mercy

(50) *When We said unto the angels, "Prostrate before Adam," they prostrated, save Iblīs. He was of the jinn and he deviated from the command of his Lord. Will you then take him and his progeny as protectors apart from Me, though they are an enemy unto you? How evil an exchange for the wrongdoers!* (51) *I did not make them witnesses to the creation of the heavens and the earth, nor to their own creation. And I take not those who lead astray as a support.* (52) *On the Day when He says, "Call those whom you claimed as My partners," they will call upon them, but they will not respond to them, and We will place a gulf between them.* (53) *The guilty will see the Fire, and know they shall fall into it, but they will find no means of escape therefrom.* (54) *And indeed We have employed every kind of parable for mankind in this Quran. And man is the most contentious of beings.* (55) *And naught prevents men from believing when guidance comes unto them, and from seeking forgiveness of their Lord, save that [they await] the wont of those of old to come upon them, or the punishment to come upon them face-to-face.* (56) *And We send not the Messengers, save as bearers of glad tidings and as warners. And those who disbelieve dispute falsely in order to refute the truth thereby. They take My signs and that whereof they were warned in mockery.* (57) *And who does greater wrong than one who has been reminded of the signs of his Lord, then turns away from them and forgets that which his hands have sent forth? Surely We have placed coverings over their hearts, such that they understand it not, and in their ears a deafness. Even if thou callest them to guidance, they will never be rightly guided.* (58) *And thy Lord is Forgiving, Possessed of Mercy. Were He to take them to task for that which they have earned, He would have hastened the punishment for them. Nay, but theirs is a tryst, beyond which they shall find no refuge.* (59) *And those towns, We destroyed them for the wrong they did, and We set a tryst for their destruction.* (60) *And when Moses said unto his young man (fatā), "I shall continue on till I reach the junction of the two seas, even if I journey for a long time."* (61) *Then when they reached the junction of the two, they forgot their fish, and it made its way to the sea, burrowing away.* (62) *Then when they had passed beyond, he said to his young man (fata), "Bring us our meal. We have certainly met with weariness on this journey of ours."* (63) *He said, "Didst thou see? When we took refuge at the rock, indeed I forgot the fish—and naught made me neglect to mention it, save Satan—and it made its way to the sea in a wondrous manner!"* (64) *He said, "That is what we were seeking!" So they turned back, retracing their steps.* (65) *There they found a servant from among Our servants whom We had granted a mercy from Us and whom We had taught knowledge from Our Presence.* (66) *Moses said unto him, "Shall I follow thee, that thou mightest teach me some of that which thou hast been taught of sound judgment?"* (67) *He said, "Truly thou wilt not be able to bear patiently with me.* (68) *And how canst thou bear patiently that which thou dost not encompass in awareness?"* (69) *He said, "Thou wilt find me patient, if Allāh wills, and I shall not disobey thee in any matter."* (70) *He said, "If thou wouldst follow me, then question me not about anything, till I make mention of it to thee."* (71) *So they went on till, when they had embarked upon a ship, he made a hole therein. He said, "Didst thou make a hole in it in order to drown its people? Thou hast done a monstrous thing!"* (72) *He said, "Did I not say unto thee that thou wouldst not be able to bear patiently with me?"* (73) *He said, "Take me not to task for having forgotten, nor make me suffer much hardship on account of what I have done."* (74) *So they went on till they met a young boy, and he slew him. He*

said, "Didst thou slay a pure soul who had slain no other soul? Thou hast certainly done a terrible thing!" (75) *He said, "Did I not say unto thee that thou wouldst not be able to bear patiently with me?"* (76) *He said, "If I question thee concerning aught after this, then keep my company no more. Thou hast attained sufficient excuse from me."* (77) *So they went on till they came upon the people of a town and sought food from them. But they refused to show them any hospitality. Then they found therein a wall that was about to fall down; so he set it up straight. He said, "Hadst thou willed, thou couldst have taken a wage for it."* (78) *He said, "This is the parting between thee and me. I shall inform thee of the meaning of that which thou couldst not bear patiently."* [al-Kahf, XVIII: 50-78]

1. And When (*Wa Idh*)

The manner in which the story of Moses with al-Khiḍr begins in Surat al-Kahf catches our attention, relating to a hidden sign. The Arabs were very sensitive to every Arabic word and the way it is used. They use the word *idh* (when) to express action that happens at a particular point in time. It is similar to using the word *ḥīna* (when). *Idh* is not used for a slow, gradual action, but rather for a sudden action or to point to a rapid change from one state to another. The word *wa* (and) is one of the most commonly used conjunctions to express parallelism. It requires a parallel between what comes before and after it. Therefore, using the word *wa* at the beginning of a segment that is not related to what preceded it, followed by the word *idh*, is an unusual usage of the two words. Using *wa idh* (and when) in the Quran and at the beginning of a segment or account is, to the best of my knowledge, the first instance of this usage in the history of the Arabic language.

After this singular linguistic introduction (*wa idh*), the story begins with no prior warnings or introductions, as if we were continuing a story or an account that had been interrupted. It seems as if this account is not related at all to what preceded it.

After the first reading of Surat al-Kahf, we may pause for a moment to notice these sudden shifts from one subject to the other and think that they apparently have nothing in common. However, an attentive reading draws our attention to the fact that the previous segment, preceding the story of Moses and al-Khiḍr, also begins with the same words *wa idh* without any clear logical, linguistic justification, according to our reasoning. This occurs in verse 50: *And when We said unto the angels, "Prostrate before Adam..."*

If we need to go back and look for the same words (*wa idh*) elsewhere, we find them in Surat al-Baqarah, at the beginning of the account when Allāh is informing the angels of His Will to make a vicegerent on earth: *And when thy Lord said to the Angels, "I am placing a vicegerent upon the earth,"...* [al-Baqarah, II: 30]

After this verse, we read the conversation between Allāh and the angels. This exchange is well known to commentators and knowledgeable people because of its peculiarity; then we read about Allāh teaching Adam all the names and Adam informing the angels of those names. At that moment the divine order is given to the angels to prostrate before Adam with the same words used in Surat al-Kahf, notably with the use of *wa idh*. After that, we find the story of Iblīs refusing to prostrate and how he induced Adam to commit sin. Right after that comes a long segment about the Children of Israel in which *wa idh* is used repeatedly, especially in verses that were meant to remind them of their deeds and their past.

This passage about the Children of Israel begins with Allāh saying: *O Children of Israel! Remember (udhkurū)...* [al-Baqarah, II: 40], then the same sentence is repeated for the second time at the beginning of verse 47, as if this repetition were a further emphasis on *O Children of Israel! Remember...* The passage continues by reminding them of their past with verses that start from time to time with *wa idh* instead of *wa-dhkurū* (and remember), as if it were a contraction of the verb. It is interesting to note that the three letters forming the expression *wa idh* are the same three first letters in the Arabic verb *wa-dhkur* (and remember), as used repeatedly in Surat Maryam, for example. These different occurrences of the expression *wa idh* show us that it is used in the Quran **as a sign, inviting readers to remember.**

This conclusion has been confirmed by many authorities, of which we refer to two sources. One of them is one of the most authoritative linguists of the Arabic language, the other is one of the most authoritative commentators of the Quran. In *Mughnī al-Labīb*, Ibn Hishām al-Ansārī says that when *wa idh* occurs at the beginning of stories related in the Quran, most of the time it is the equivalent of the imperative of the verb meaning remember (*idhkur*). Ibn Hishām's opinion concerning *wa idh* is confirmed by al-Suyūṭī in his book *Al-Itqān fī 'Ulūm al-Qur'ān.*

Mindful, attentive readers of the Quran with a sharp memory would certainly connect similar passages to reach an accurate understanding of a given message that is an integral part of the holistic message which is the Holy Quran. In fact, every passage of the Quran should be approached and understood as a commentary, explanation, example and conclusion of a former passage and a preparation for a subsequent one.

It becomes an even easier mission when each passage of the Quran is considered part of a set of signs and part of a means, which serve one precise goal: a message directed to mankind. At the same time, it is part of a universal fabric in which human beings are invited to participate. Therefore, this *wa idh* starting the story of Moses and al-Khiḍr leads us to link it with the preceding section that begins: *And when We said to the angels...*, to the details that follow about Iblīs refusing to prostrate before Adam, and the resulting dramatic consequences.

In these verses of Surat al-Kahf, from *And when We said to the angels...* to *but they will find no means of escape therefrom*, we have a very concise, dramatic summary, rephrasing a subject that is presented in detail in many other places in the Quran. Each time the subject is presented, it is done in a way that suits the general context and emphasizes a particular idea.

2. Knowledge

Wa idh is an invitation to remember the blessings that Allāh has bestowed upon us, namely His true revelation; so we should accept the invitation *So let them respond to Me* [al-Baqarah, II: 186], for truly the remembrance benefits the believers (cf. al-Dhāriyāt, LI: 55).

This *wa idh* led us to link the section about Moses and al-Khiḍr with the section that preceded it and begins with: *And when We said to the angels...* This then refers us to the previous occurrence of the expression and traces it to the first time it was used, which was in Surat al-Baqarah, where we find the first account of the story of Adam in the Quran, in other words, the point where everything started for us. Knowledge requires us to be aware of how things started in order to be able to understand how they end.

Given that this is the first time this account is presented in the Quran, we can suppose that the details which it contains are particularly important. Allāh said: *And when thy Lord said to the angels, "I am placing a vicegerent upon the earth," they said, "Wilt Thou place therein one who will work corruption therein, and shed blood...?"* [al-Baqarah, II: 30]. It is likely that anyone who has read the Quran must have wondered: how did the angels know? This is a normal reaction expected of the readers, especially using the word "know" (*ya'lamū*). The Quran states: *"... (one who will) shed blood, while we hymn Thy praise and call Thee Holy." He said, "Truly I know (a'lamu) what you know (ta'lamūn) not."* [al-Baqarah, II: 30]

The last word used is *ta'lamūn* (you know). The root of this word (*maṣdar*) is the verb *'alima* (to know) from which derives the noun *'ilm* (knowledge), which is written exactly like *'allama* (to teach). The following verse begins with the verb *'allama* (He taught). If we add the letter *ta* to it, it becomes *ta'allama* (to learn). Translated into any other language, English for example, we need a different verb for each of these forms and each has a different root, whereas in Arabic, we use the same root *'alima* for all of these variations which all contain the same three letters (*'a la ma*). We are often oblivious of the importance of this unique feature of the Arabic language due to the veil of our familiarity. Therefore, what is in fact a stressed emphasis on a word which has a very deep significance easily gets lost in any translation.

In verse 30: *...lā ta'lamūn* (*...you know not*) and in verse 31: *'allama...* (*And He taught...*) the emphasis is clearly on knowledge. This emphasis was introduced by the last word [*'alīm*] in the section that precedes the first *wa idh* in which Allāh says: *... and He is Knower of all things ('alīm) And when thy Lord said....* [al-Baqarah, II: 29-30]

And when (wa idh) thy Lord said...

Should we not have a closer look at our understanding of the word 'said' when it is related to Allāh? Our understanding of any word is shaped by our previous encounters with it and all the ways it has affected our emotions, imagination, impressions, beliefs, and memories. Can we extrapolate this general worldly conception onto the Divine? Or do we have to enlighten our understanding of the word as much as possible so as to come closer to the meaning and take a further step on the path of knowledge, out of fear of Allāh's words: *They did not measure Allāh with His true measure?*[19]

And when (wa idh) thy Lord said...

Why is 'thy Lord' used? Why not 'Allāh said...,' or 'the Creator said,' or 'the Lord of the Worlds said'? Why was the first verse revealed to the Prophet Muḥammad ﷺ: *Recite in the Name of thy Lord...* rather than *"Recite in the Name of Allāh"*? It is a legitimate question given that we are dealing with a text in which verses and words were perfectly composed, not by a brilliant writer or skillful thinker, but by the Creator of the Universe and of all existing things. It is a text in which He included a message to lead us from darkness into light.

And when (wa idh) thy Lord said to the angels...

Is this statement meant as useless information, a mere conversation with the angels about the matter in question? Or is it a notification with a particular purpose and an assigned mission for that created being (Adam), as it appears in Allāh's command to the angels to prostrate before Adam?

And when (wa idh) thy Lord said to the angels "I am placing..."

Allāh did not say: "I am creating."

And when (wa idh) thy Lord said to the angels "I am placing upon the earth..."

'Upon the earth' is a detail we rarely stop at, because our attention is caught by the rest of the verse. In this expression resides the essential difference between the Quran and the published versions of the Scripture of the People of the Book (the Old and New Testament). This is a major difference between the Islamic and the Christian versions of the Fall. The foundation of Christian doctrine is Original Sin, from which the only way to salvation is through the Messiah. The idea, as it appears in the first pages of Genesis, is as follows: after Allāh created all of creation, He created Adam and gave him a wife and placed them in the Garden of Eden and ordered them not to eat from the forbidden tree. However, Satan persuaded Adam to sin, by his wife enticing him, and that is how the sin was committed. The Biblical account gives the reader the impression that Allāh's Will was contradicted by the sin of Adam and that Satan's ruse ruined everything, changing the course of events and contradicting Allāh's happy plan. The conclusion therefore is that Satan was stronger than Allāh! (Glorified be He). This idea is confirmed by the account of the Flood. According to the Biblical version, when Allāh saw that human wickedness was spreading throughout the earth, He decided to wipe out all mankind. It is explicitly expressed that 'God was grieved' for having created mankind![20] Exalted is He above that which they ascribe.

19. [Editor's footnote] See al-Ḥajj, XXII: 74; al-An'ām, VI: 91; and al-Zumar, XXXIX: 67.

20. Chapter 6 of Genesis: "(5) And God saw that the wickedness of man was great in the earth, and that every thought in his heart was only continual evil. (6) And it repented the Lord that he had made man on the earth, and it grieved Him in His Heart. (7) And the Lord said, I will destroy man whom I have created from the face of the earth; both man, and beast, and the creeping things, and the fowls of the air; for it repenteth Me that I have made them." [Editor's comment]: Throughout the book, the

Verse 30 of Surat al-Baqarah presents a version that is completely different and transcends the problem we described above. That is because when the story of Adam is related for the first time in the Quran, it does not declare 'I am placing in paradise,' but: *"I am placing upon the earth."* It means that Allāh originally intended Adam to be on earth and this did not come as the result of an unexpected situation.

Upon the earth: why was the earth chosen rather than any other place in this vast universe? This was decided upon through the Wisdom of the Creator of the earth which He alone possesses. We might catch a glimpse of this wisdom if we stop and look at the main characteristics of the earth.

If we cast away the veil of familiarity that blinds our inner sight, we will be able to see that the main feature of the earth is harm. Even if spared all the numerous harmful conditions and catastrophes, anything left on earth, whatever beauty it may have, will die and will be transformed into a repulsive cadaver. On this earth, there are countless sources and causes of harm. Nothing stays firm before the power of decay, even the hardest rocks. Every prophet, saint or man of Allāh has suffered pains that only the most forbearing could sustain. Every time a prophet educated a generation of righteous people, there followed another generation of people who lowered themselves to a lower state. Even after the return of Jesus, peace be upon him, when he restores good on earth for a short period of time, the souls of the righteous will be taken and only the worst kinds of people will dwell on earth until the Day of Judgment. Is this trait of harm, which is specific to the earth, a result of the deeds of creatures, or is it linked to the Will of the Overseer of Creation (*al-Muhaymin*) – *for thy Lord inspired her [the earth]* [al-Zalzalah XCIX: 5] – the Will of Him, Who has a hidden Wisdom, which is a manifestation of one of His Names (may His Glory be exalted)?

Was the harmful feature of the earth unknown to the angels (peace be upon them)? It is clear from the Quranic verse that they were aware of it and this is what confused them. Let us consider an example, although Allāh is above any form of comparison: a king wishes to choose some of his subjects for a particular position; because he is just and fair, he wants to have evidence which proves them worthy of being placed in that position. At the same time this evidence will prove that the unworthy do not deserve to occupy that position. The king informs one who is obedient and close to him that he would send the subjects to a city known for its polluted air, bad location and food, a place where corruption and bloodshed prevail. His advisors are confused and ask the king why he would send them to this city and let them become worse than its inhabitants, influenced by the bad atmosphere of the city? Exalted is He above any form of comparison; this is only a crude approximation.

We can then imagine the noble nature of those who can resist the negative influence, break through it, and then, instead of descending, rise to higher ranks. This is the meaning of the passage through this life and all the trials for the

word Allāh is used instead of 'God' based on the author's preference, see *Kitāb al-Iḥsān* by Suleiman Joukhadar, Dār al-Fikr, Damascus, 2016; the English translation is forthcoming.

servants who are challenged to show the quality of their true, inner selves, and all that they are hiding: *...to witness their deeds.* [al-Zalzalah, XCIX: 6]

It also shows how naïve it is for anyone to dream of building a paradise on earth and to live there eternally, when in fact it is predestined to vanish. Has corruption not appeared on land and sea? (cf. al-Rūm, XXX: 41) It shows the ignorance and short-sightedness of people who choose to reside in a world that was created as a passage, without knowing what vessel they are in, the reason they are in it and where it is taking them. How can someone who is heedless of this foundation claim to build knowledge?

> *And when (wa idh) thy Lord said to the angels, "I am placing upon the earth a vicegerent"...*

3. Adam and Knowledge

Because we are confronted with the first Quranic account of the story of Adam, and after reading about the Will of Allāh to place a vicegerent upon the earth, we may expect with some naivety to be given details about the creation of Adam out of clay. However, this part of the story is not presented here. This means that the fact that Adam was created from clay is secondary, compared to another detail that was given priority over anything else, namely knowledge: *"...you know not." And He taught Adam the names, all of them....* [al-Baqarah, II: 30-31]

It is very interesting to note how verse 30 ends with ignorance, *know not*, and verse 31 begins with teaching, *He taught*. Not teaching the angels, but teaching our Master Adam. The first time the name Adam appears in the Quran it is followed by the word *the names*. What is meant by *all of them* remains a mystery, even though all commentators have tried to provide an interpretation for this. Some of them did not even realize that this is a mystery. Only those who were given tidings of His Mercy and received His Knowledge know what is meant by *all of them*.

> (31) *And He taught Adam the names, all of them. Then He laid them before the angels and said, "Tell me the names of these, if you are truthful."* (32) *They said, "Glory be to Thee! We have no knowledge save what Thou hast taught us. Truly Thou art the Knower, the Wise."* (33) *He said, "Adam, tell them their names." And when he had told them their names He said, "Did I not say to you that I know the unseen of the heavens and the earth, and that I know what you disclose and what you used to conceal?"* [al-Baqarah, II: 31-33]

In verse 29: *the Knower*, in verse 30: *...I know... you know not*, in verse 31: *And He taught...*, in verse 32: *...no knowledge... Thou hast taught us... the Knower*, and in verse 33: *...I know... and that I know...*

All these words that are derived from the root (*maṣdar*) *ʿalima* are used 9 times in the five verses: 29, 30, 31, 32, 33. This is eye-catching. If we stop and think about the information given in this allusion, we discover that it is one of many steps towards understanding who is intended by the pronoun used in the following verses:

> (31) *Then He laid them...*

> (34) *And when We said to the angels, "Prostrate unto Adam," they prostrated, save Iblīs. He refused and waxed arrogant, and was among the disbelievers.*

In the Quran, the same sentence is repeated 5 times with the same words: *...And when We said to the angels, "Prostrate unto Adam," they prostrated, save Iblīs...* In al-Baqarah, II: 34; al-Isrā', XVII: 61; al-Kahf, XVIII: 50; Ṭāhā, XX: 116; and al-Aʿrāf, VII: 11. 4 times as: *And when We said to the angels*, and once *then We said unto the angels*. 4 + 1 = 5. This is a wonderful allusion.

We find the same divine command to the angels to prostrate before Adam two more times, with another wording: *fall down before him prostrating*. In both cases (in al-Hijr, XV: 29 and Ṣād, XXXVIII: 72) the same words are used: *"When I have proportioned him and breathed into him of My Spirit, fall down before him prostrating."* In these two verses, Allāh did not say before Adam. He said: before him. The pronoun *him* does not refer to Adam but to the human being (*bashar*) which appears in the verse before it (and which of course refers to Adam *in extenso*). This means: 7 times = 5 times prostrating before Adam as mentioned by name and 2 times prostrating before him as a human being. 5 + 2 = 7. This is a wonderful allusion.

What is important here is that the divine command to prostrate before Adam appears in Surat al-Baqarah immediately after the verses about the knowledge of names, after Allāh taught Adam all the names. In the other surahs of the Quran, al-Hijr [15]: 29 and Ṣād [38]: 72, the command to prostrate comes immediately after Allāh's breathing of His Spirit into Adam. The verses in both surahs are very similar in words and in meaning: "*...so when I have proportioned him and breathed into him of My Spirit, fall down before him prostrating.*"

Therefore, the command to prostrate before Adam as a human being came after Allāh breathed His Spirit into him and taught him divine knowledge, which is the knowledge of names. Do I still need to put further emphasis on this point? The Spirit and knowledge were the gifts to Adam as a human being.

4. Iblīs and Ignorance

The angels obeyed Allāh's command, whereas at the same moment Iblīs disobeyed and refused to prostrate before Adam/the human being who was given of the Spirit of Allāh and knowledge of the names.

In each one of these seven accounts of the story of Adam we notice that the words which come right after the name of Iblīs undergo change. This change emphasizes an aspect of Iblīs's satanic nature, in accordance with the message conveyed in every account and the general contexts.

Because we now have many verses about the same subject in front of us, we may very well compare them. As a first step, the correct and simple way to approach these details is by looking at the similarities, then the differences in them. It is not possible to understand a subject without knowing its beginning and its basis. It is therefore correct and natural for human beings to wonder about their origin, their existence, their destiny, how the human drama began. This is one of the greatest questions ever asked.

The necessary information to answer these questions is provided in the Quran, while insisting on the fact that Iblīs is the main enemy of mankind. This information also teaches human beings how to protect themselves. The best way to protect oneself is to know the nature of the enemy, his purposes, means and ways.

The seven verses mentioned above have in common Allāh Almighty's command to the angels to prostrate before Adam, and they all did so, except Iblīs. In the seven verses mentioned, Iblīs is set apart from the angels.

Verses 30 and 31 of Surat al-Hijr (XV), and verses 73 and 74 of Surat Ṣād (XXXVIII) seem to confirm this idea: *Thereupon the angels prostrated, all of them together, save Iblīs. He refused to be with those who prostrated.*

Because the command was given to the angels and Iblīs disobeyed Allāh, readers might at first think that Iblīs was an angel before the divine command was expressed. 7 - 1 = 6 verses lead us to believe this. But none of these 6 verses provides us with clear information about the nature of Iblīs. The seventh verse, however, is clear and explicit. Verse 50 of Surat al-Kahf provides clear and explicit confirmation that Iblīs was of the jinn: *...they prostrated, save Iblīs. He was of the jinn and he deviated from the command of his Lord.*

The fact that we have six verses which make us believe (because of the Biblical influence) that Iblīs was an angel, and that we have a seventh verse which explicitly states that he was of the jinn, compels us to stop and think about this contradiction. To pay attention to the meaning understood from the six verses, because of their number, and to ignore the seventh, assuming it to be secondary, is incorrect and lacks insight. Nevertheless, there are some who are of this opinion. That is because they have fallen into a trap when they confuse Iblīs with the angels in the other six verses.

The confusion occurs because they sought the necessary information in the wrong place.

These verses have specific purposes and different goals, but they do not include explicit tidings about the nature of Iblīs. Confronted with the seven verses, those who confused Iblīs with the angels saw that the solution was to choose verses which do not contain anything about the nature of Iblīs as a reference and to give the clear and explicit seventh verse the same meaning as the other six. According to their reading, Iblīs was not of the jinn, but became a jinn.

The verb *kāna* (to be) has a secondary usage, meaning: to become. Their arguments are that in the seven verses the meaning of the verb 'to be' changes to this secondary meaning. 'To be' cannot mean 'to become' in the past, in the case of a complete change in his origin and nature.

When did Iblīs become one of the jinn? According to these commentators, after he disobeyed Allāh, he was cursed and fell from the rank of angels to that of a satanic nature, thus becoming the first jinn to exist. However, it is well known that angels were created out of light. Iblīs himself declared that he was created out of fire: *He said, "I am better than him. Thou hast created me from fire, while Thou hast created him from clay."* [Ṣād, XXXVIII: 76]

These commentators might object by saying nothing proves that all angels were exclusively created out of light. They maintain that Iblīs was an angel who was created from fire.

In order to have an accurate reading of the knowledge provided in the Quran, a comparison with other information related to the same subject is needed. The fact that we have 5 + 2 = 7 makes the comparison easier. 7 - 5 = 2 verses (in Surat al-Ḥijr and Surat Ṣād) which are very similar. In Surat Ṣād (XXXVIII), Iblīs declares that he was created out of fire. The verses that come right before verse 29 in Surat al-Ḥijr (XV) provide us with details and an answer to an expected question concerning Iblīs's nature:

> (26) *And We indeed created man from dried clay, made of molded mud* (27) *and the jinn We created earlier from scorching fire* (28) *And [remember] when thy Lord said unto the angels...*

The contrast between man and jinn is clear: jinn = fire + air; man = earth + water.

Because that detail is provided immediately before one of the 5 + 2 = 7 verses studied above, we have a verse in which Iblīs is comparing his origin from fire to the origin of man from earth. We have a clear and explicit verse informing us of the origin of Iblīs [al-Kahf, XVIII: 50] and there is no other indication as to his nature. The opinion, maintaining that Iblīs was cursed and his punishment was to be lowered from the rank of angels to that of a jinn, is not acceptable.

Verse 50 in Surat al-Kahf: *...save Iblīs. He was of the jinn and he deviated from the command of his Lord...* clearly states that Iblīs was of the jinn before the divine command to prostrate. The other verses state that the jinn were made from fire before Adam was created. Therefore, there should not be any confusion between angels and Iblīs. Sīdī al-Ḥasan al-Baṣrī was categorical as far as this question is concerned when he declared: "Iblīs was never an angel; not even for one second" (see Ibn Kathīr's commentary on the Quran).

One problem remains unresolved however: the divine command to prostrate before Adam was given to the angels. Iblīs was of the jinn, meaning that the command was not of concern to him. So why should he comply with a command that was not addressed to him?

The answer is simple and clear for anyone who is aware of the ancient, universal manners, which were obligatory. At times when different generations of the same family meet for a social occasion, if someone who belongs to a 'higher rank' within the family shows humility and exaggerated respect towards another person, anyone of a lesser rank would inevitably have to do at least the same, in accordance with the laws of courtesy and good manners. Allāh's command was given to the angels, who are of a higher rank compared to other creatures. Therefore, anyone enjoying a lesser status who heard the command would have been obliged to obey and prostrate, including Iblīs.

> **Are these facts unclear in the Quran? Definitely not. What is unclear is the mind of anyone who does not see these obvious facts.**

In the previous case presented above, we dealt with 7 elements of knowledge. The fact that it is 5 + 2 = 7 is an allusion that we can understand in a profound manner through the science of numbers. However, this approach is higher than can be presented in this introduction.

A meticulous study of the 5 + 2 = 7 elements, based on the next example, is complicated. To give a simplified example of a primary approach to this knowledge, let us reduce these 5 + 2 = 7 elements to 7. Let us consider them as 7 points on a circle, constituting the vertices of a seven pointed star, a heptagram.

We may complete the regular construction of a seven sided shape, or heptagram by tracing a line through each of the seven dots, provided that each angle is 360/7 degrees. In this way we trace a perfect heptagram. If any of the seven dots on the circle is omitted, and we have 7 - 1 = 6 dots, by removing a specific point we move to angles that are 360/6 degrees. Then the shape will have 6 sides and none of the points will be identical to any of the 7 points of the heptagram drawn on the circle. **There is a subtle allusion in this example.** The Quran should be approached in this same meticulous way.

We saw that considering six elements and leaving out the seventh leads us to an erroneous result. We may therefore conclude that one of the purposes of the Quran is not only to present us with pure elements of knowledge, but also to teach us how to deal with these elements. In the case we studied above, in the seven verses, the information we were in need of from the Quran was not presented to us once, in a single instance, and in one section, but in various ways and in different places. It is very legitimate to wonder about the purpose of such a means of transmission. The main purpose has been expressed in different ways:

> **The purpose is to lead mankind from tenebrous darkness to light; light is manifested through knowledge.**

The message of the Quran is not merely to provide theoretical elements of knowledge. This is not enough to lead mankind from darkness to light. Through the Quran we were given a way, a means, and a method to use to deal with these elements. If the method had been explicitly outlined in the text, it would then have been only a new addition to other theoretical elements pertaining to previous knowledge.

How to deal with the elements of knowledge was given in a way that enables readers to interact with the method while discovering it at the same time. The information presented in the Quran obliges readers to make an effort to comprehend it. It compels them to gather various pieces of knowledge and compare them, to carry out an accurate analysis, and then to gather the pieces together again and reach an accurate synthesis. The Quranic text elevates readers' thinking capacities from the numbness of daily life to the level of the universal. The text never lowers itself to the level of an inattentive reader.

In the case of the seven verses, the information we needed in the Quran is not presented once and in one section, but in various ways and in different places. In addition, we saw that it is presented in a seemingly confusing way, if readers are not attentive.

This is the best way to awaken the readers of the Quran from the slumber of daily thinking, to draw their attention and push them to interact with the subject. It is meant as an exercise for the intellect and to elevate their abilities, thus giving them the capacity to see clearly. How needful are they of clearly seeing and recognizing a relentless, deceitful and misleading enemy!

I must pause here before we continue with the subject we are discussing to point out that nothing in the Holy Quran suggests that Iblīs is a superior creature through his knowledge, intelligence, personality or rank. Rather, an accurate and objective reading of the Quranic text emphasizes that it is the exact opposite which is the case. Does this mean that we should not pay heed to this enemy whose schemes are weak and whose inanity testifies to his nature? On the contrary, what a great shame to fall into the traps of such an enemy: *...save Iblīs. He was of the jinn and he deviated from the command of his Lord...* [al-Kahf, XVIII: 50]

Iblīs did not obey the divine command: *...He refused and waxed arrogant....* [al-Baqarah, II: 34] This moment is precisely the origin of all evil. It is thus important to pause and reflect. In order to understand what happened at this moment, it is necessary to analyze the event and deconstruct its constituent parts, based on the following:

- The self is not moved (in a conscious way) if everything is going according to what it is used to and is seeking.
- Individuals are forcefully conscious of their self when their will is contradicted or when they are asked to do something unexpected, thus pushing them to change what they are accustomed to. This requires them to focus their will and fervor on performing the new task at hand, which is only possible by exerting control over the scattered self. The confrontation and friction between the self on the one hand and the task to be done on the other causes the individual to become aware of his self.

When Iblīs heard the command to prostrate, his *self* reacted according to the explanation above. It was moved and focused, because it was confronted with a task that needed to be carried out. The confrontation was between Iblīs's *self* on the one hand and the command to prostrate to Adam on the other.

Let us stop the wheel of events now between the moment the command was issued and the moment Iblīs disobeyed. All evil rests in this moment. **What happened?** Because the command was so unusual for Iblīs, compared to what he believed and accepted, the process explained above was triggered, resulting in an extreme polarization: his *self* on the one hand and his prostrating before Adam on the other. The contrast was so extreme that it made Iblīs's sense of his own *self* increase noticeably. This is exactly where we reach an important crossroad.

> **On the one hand, remembering Allāh, or rather being aware of the existence of Allāh and the implications of this awareness, and on the other hand, being aware of one's self or ego.**

As far as Iblīs is concerned, his awareness of Allāh's presence was completely absent and eclipsed by his awareness of his own self. His disobedience would not have happened, were it not for his weak and frail awareness of Allāh's presence compared to his strong ego. **This is the moment that marked the beginning of the fall.**

Iblīs's ego was so strong that it completely dominated him and his capacity for judgment. His inner eye was blinded and veiled from the light of Truth. He thus deprived himself of the true point of reference, and made his own self the point of reference. His judgment was suddenly based on his self and its needs. Thus his conclusions were in conformity with his self and what pleases it. Should we not learn from this lesson?

If we look at the knowledge which Allāh has so generously given us about Iblīs, and His question to him about the reason he refused to obey Allāh, we find Iblīs's answer clearly in accordance with what we explained above. The first word Iblīs uttered, in answering Allāh's question, was *I...* He put himself before anything else and thus preferred his *ego.* He continued speaking about it and could not perceive anything else nor diverge from the subject of his *ego.* He glorified himself by saying *I am better...* and was too proud even to mention the name of his opponent, the very name that was linked to the command to prostrate, and said: *...than him...* before going back to speaking about himself and presenting other evidence that was in fact against him. He had to be completely veiled and incredibly blind to speak in this way to his Creator, while recognizing Him as his Creator: *...Thou hast created me...*

It was so foolish of him to want to present his case to prove to his Creator that He was wrong about him, praised be He! That was because he seemed to try to draw the attention of the Creator of fire and clay to the question: "Why are You not aware that fire is better than clay?" This embodied the ignorance of Iblīs:

- Iblīs thought that he had more knowledge than the All-Knowing and thought he was better placed to draw his Creator's attention to something He had forgotten, glorified be He!

- He disbelieved in Allāh, the Judge, the Wise, the Just, the True, because his refusal meant that the act of prostrating seemed unfair to him.
- He failed to make Truth a criterion for his judgment and referred instead to his own self, making it the criterion and point of reference and arguing like one who is forgetful. Fire is not superior to clay. The truth is, fire and clay are alike. Both were created by Allāh, Who put into each element many of His wonders and secrets.

Should we not learn from this lesson?

The reason Iblīs was pushed in his stubborn disobedience, while the angels prostrated, is that he was completely immersed in the darkness of his own *self* and the objectives he was seeking. From his point of view and according to his reasoning, refusing to prostrate, while the angels prostrated, signified that he was (in some manner) set apart, thus superior to them and at a higher rank compared to them. And given the high rank of angels, his refusal made him even more distinguished than them in the blink of an eye. This is what his *self* was desperate for, in order to escape any kind of authority. Iblīs's plight resides in that authority, because the angels in fact have authority and power over the jinn.

Iblīs forgot that the power in any created being can only originate with the Creator and only takes effect according to His Will. What motivated Iblīs's ambition and wish to be better than the angels was that he feared something other than Allāh Almighty. It was the fear of feeling inferior to a more powerful creature, a fear of the other creature's authority and the need for authority for himself.

Should we not learn from this?

Iblīs used this feeling of inferiority and wanted to impose his own plight upon his opponents, Adam and his wife. His plan was:

C. To make them feel inferior, so they would feel a need to compensate for this shortcoming, which would be enough to drive them to commit the sin

B. To weaken their resistance by shaking their faith in Allāh – the source of the light that enlightens the individual's insight – which was necessary for this to happen

A. To divert their attention to their own selves so they would be veiled from seeing the Truth, which was necessary for him to succeed in this.

So this is what he did:

A. *Thus he lured them on through deception...* [al-A'rāf, VII: 22]. Iblīs succeeded by proceeding gradually. He began by making Adam aware of his *self* when he called him *...O Adam!...* This was the only time that Iblīs put aside his pride and called his opponent by his name, but he only did it to make Adam fall into the veil of the *self*. As a result of this call, Adam became aware of his *self* and turned his attention to the caller. Adam was thus ready for the rest of the enticement of Iblīs, who exploited mankind's great natural love for good:[21] *...Shall I show thee the Tree of Everlastingness and a kingdom that never decays?* [Ṭāhā, XX: 120]

21. [Editor's footnote] See al-'Ādiyāt, C, 8

Iblīs did not oblige Adam to do anything, but made him feel a need in his *self* through a question that had just two possible answers. 'Yes' was surely the inevitable answer.

The idea of the Tree of Everlasting Life dominated Adam's imagination and the desire to reach it was so strong that it became difficult for him to give it up in the face of the Truth. Indeed, it requires a great effort to overcome one's desires for the sake of obeying Allāh. *And We indeed made a pact with Adam aforetime, but he forgot. And We found no resoluteness in him.* [Ṭāhā, XX: 115]

B. The situation now required severing the relationship between Adam and the Truth, which was achieved by weakening his faith by means of planting misleading and incorrect ideas: *"...Your Lord has only forbidden you this tree, lest you should become angels, or among those who abide [forever]."* [al-A'rāf, VII: 20]

By this, Iblīs wanted to inspire Adam and his wife to think that:

- Their Lord was not truthful concerning the reason for forbidding the tree. If that were true, then it would be legitimate to doubt that Iblīs is the enemy. In this way, Iblīs is well placed to claim: *"...Truly I am a sincere adviser unto you"* [al-A'rāf, VII: 22] and to present an alternative reason for the prohibition. This doubt of Allāh's Truthfulness is ignorance. *And who is truer than Allāh in speech?* [al-Nisā', IV: 87]
- Their Lord forbids goodness out of meanness. This is ignorance, similar to that of the Children of Israel when they said *"Allāh's Hand is shackled."* [al-Mā'idah, V: 64]; it is ignorance of Allāh's boundless generosity.
- Their Lord did not wish them to have goodness, so why place one's hope in Him? This is ignorance of the fact that Allāh is kind (*Laṭīf*) unto His servants.
- Their Lord is wrong in His prohibition, meaning He is not Just. This is ignorance of the fact that Allāh is the Most Just Judge, the One Who masters proportions of all things.
- Their Lord is an opponent, because He is interfering with their ambitions, so why obey Him? This is ignorance of Allāh's Omniscience: it is real ignorance to consider that the created and the Creator are alike and in confrontation as opponents. It is similar to the contrast people make when they compare Allāh and Satan: the way in which they see absolute Good and absolute Evil. This is utter ignorance because nothing can be compared to Him, praised be He.
- The tree offers Everlasting Life and a kingdom that never perishes and not their Lord. This is ignorance that all power is in fact in the Hands of Allāh, the sole Almighty.
- If they eat from the tree they overcome the will of their Lord and impose the reality of things on Him. So why fear Him? This is ignorance – that nothing happens without Allāh knowing it nor is it outside of His Will.

C. This is how Iblīs managed to entice Adam and his wife into the trap and impose his plight on them. He made them believe that:

- They are not immortal; i.e. they are destined to pass away while their utmost wish is to enjoy immortality.
- They do not have a kingdom that never perishes; they are subject to being in need at every moment.

- They are not angels, so they may become elevated to a higher rank by becoming angels. As long as this has not happened, they are inferior.

By diverting Adam's and his wife's attention from Allāh to their own *selves*, Iblīs managed, through his calling, to give them illusions which had an effect on their *selves* because they surrendered to them. These illusions created in their selves a need that they sought to fulfill through their own powers. They were moved by their own illusions, with no coercion from Satan.

I firmly believe that every detail mentioned in the Quran is absolutely indispensable. This brief analysis proves that the details which Allāh so generously provides in His Book are not meant as embellishment for the story recounted, or for entertainment. They are pauses in which one is invited to reflect, draw conclusions, and learn from the lessons provided. Can one who is seeking true knowledge drink from the sources of knowledge, science, truth and guidance if the doors of ignorance and its sources have not been closed to his *self*?

5. Iblīs's Errors

And when (wa idh)... We said to the angels, "Prostrate unto Adam," they prostrated, save Iblīs. He refused and waxed arrogant, and was among the disbelievers. [al-Baqarah, II: 34]

Iblīs committed many grave errors:

1. Iblīs was not aware of his true limits, thus refusing to obey and falling into arrogance, based on his erroneous judgment of who he was.
2. *...(he) was among the disbelievers*: his ignorance turned him away from faith. He turned away from Truth and held to his own false opinion.

   Verse 75 of Surat Ṣād reads: *[Allāh] said, "O Iblīs! What has prevented thee from prostrating unto that which I created with My two Hands? ...."* Iblīs did not understand that the divine command is Absolute Truth, beyond any doubt. He did not understand that the divine command was to prostrate before what Allāh had created with His Hands and to what Allāh had offered Adam of His Spirit and His Knowledge! *[Allāh] said, "O Iblīs! What has prevented thee from prostrating unto that which I created with My two Hands? Dost thou wax arrogant, or art thou among the exalted?"*

   Verse 33 of Surat al-Baqarah reads: *Did I not say to you that I know the unseen of the heavens and the earth, and that I know what you disclose and what you used to conceal?* Subsequently in verse 75 of Surat Ṣād, Allāh did not need Iblīs to give an explanation, and most probably gave Iblīs a chance to repent and correct his error by showing him his misunderstanding and by warning him against overestimating his *self*. The question was a particular reprimand and the expected answer was meant to reveal what was hidden and also to give evidence that would be held against Iblīs, who replied: *"...I am better than him. Thou hast created me from fire, while Thou hast created him from clay."* [al-A'rāf, VII: 12]
3. Iblīs only saw the superficial and material aspect of the question from his own personal point of view: the contrast between fire and clay.
4. Based on incomplete information, on deficient knowledge and erroneous reasoning, Iblīs came to the conclusion that supported his claim.
5. Iblīs did not take the opportunity that was offered to him by the Clement, the Patient, the Judge, the Just, the True, glory be to His Majesty, in order to rectify his arrogance and stubbornness and his holding with his position based on false information and premises.

Are these errors not very familiar?

6. The Human Drama

In the preceding commentary on Surat al-Baqarah and Surat Ṣād, Iblīs's errors, ignorance and perversion were made clear to us. It is worthy of note that in Surat al-Hijr, he swears to employ his own faults to mislead mankind as he himself was misled: *He said, "My Lord! Since Thou hast caused me to err, I shall surely make things seem fair unto them on earth, and I shall cause them to err all together."* [al-Hijr, XV: 39]

If we go back to Surat al-Kahf, we will have a deeper understanding of verses 50 to 53, which start with an allusion inviting the reader to remember. This is exactly what we did in the exercise above.

> (50) *"And when* (wa idh)*... We said unto the angels, "Prostrate before Adam," they prostrated, save Iblīs. He was of the jinn and he deviated from the command of his Lord. Will you then take him and his progeny as protectors apart from Me, though they are an enemy unto you? How evil an exchange for the wrongdoers!* (51) *I did not make them witnesses to the creation of the heavens and the earth, nor to their own creation. And I take not those who lead astray as a support.* (52) *On the Day when He says, "Call those whom you claimed as My partners," they will call upon them, but they will not respond to them, and We will place a gulf between them.* (53) *The guilty will see the Fire, and know they shall fall into it but they will find no means of escape therefrom."*

These verses from Surat al-Kahf emphasize one particular point: the real human drama began with Allāh's command to the angels to prostrate before Adam. In other words, Allāh gave Adam the greatest positive power integrally linked to Spirit/Knowledge. At this particular moment, based on the universal law of equilibrium, Iblīs, integrally linked to ignorance, became the greatest negative power against Adam.

We can consider the divine command to bow down to Adam as a turning point. **That turning point made apparent what was hidden and what was veiled,** and a clear distinction between good and evil was drawn.

In these verses of Surat al-Kahf, after prostration to Adam was mentioned, we are informed of those who went astray and their subsequent punishment, from which they found no escape. The circle from creation to resurrection, as summarized in the previous verses, ends in a dramatic way. Can we forget Allāh saying: *And when their eyes turn toward the inhabitants of the Fire, they will say, "Our Lord! Place us not among the wrongdoing people!"* [al-A'rāf, VII: 47]

If we, as readers of the Quran, are sincere and sensitive and interact with the text while reading it, and are not arrogant and know that there is nothing that guarantees that we will not be with those in the Fire, then we shall ask ourselves: where is the escape (*maṣrif*) from this Fire from which the guilty wrongdoers find no escape?

7. The Escape

All we need to do is to continue reading Surat al-Kahf: *...they will find no means of escape (maṣrif) therefrom. And indeed We have employed (ṣarrafnā) every kind of parable for mankind in this Quran.* [al-Kahf, XVIII: 53-54]

> (54) *And indeed We have employed every kind of parable for mankind in this Quran. And man is the most contentious of beings.* (55) *And naught prevents men from believing when guidance comes unto them, and from seeking forgiveness of their Lord save that [they await] the wont of those of old to come upon them, or the punishment to come upon them face-to-face.* (56) *And We send not the messengers, save as bearers of glad tidings and as warners. And those who disbelieve dispute falsely in order to refute the truth thereby. They take My signs, and that whereof they were warned, in mockery.* (57) *And who does greater wrong than one who has been reminded of the signs of his Lord, then turns away from them and forgets that which his hands have sent forth? Surely We have placed coverings over their hearts, such that they understand it not, and in their ears a deafness. Even if thou callest them to guidance, they will never be rightly guided.* (58) *And thy Lord is Forgiving, Possessed of Mercy. Were He to take them to task for that which they have earned, He would have hastened the punishment for them. Nay, but theirs is a tryst, beyond which they shall find no refuge.* (59) *And those towns, We destroyed them for the wrong they did, and We set a tryst for their destruction.* [al-Kahf, XVIII: 54-59]

The main idea in these verses, based on the reasoning we saw before, revolves around man's negative reactions towards the divine message which contains man's salvation. The emphasis in the verses above is on man's misunderstanding of the divine message. So how can man avoid misunderstanding and be protected from it? Through avoiding mistakes.

And indeed We have employed every kind of parable for mankind in this Quran. [al-Kahf, XVIII: 54] The promised parable is provided: we have a story which clarifies these mistakes, especially in the case of seeking divine knowledge.

8. Our Master Moses

As part and parcel of what we have seen before, and as a commentary, explanation, and conclusion, the story of Moses and al-Khiḍr is an example of an extreme case. It is noteworthy that of all the other prophets, Moses was chosen to be part of this story, which was predestined in His Eternal Knowledge to be in the Holy Quran until the end of time. Moses is known for his character, exceptional capabilities, and his great knowledge that included sciences that were known to the Egyptian elite. Theirs could be considered the highest degree of knowledge that existed in the whole world at the time. Therefore, the story is that of an exceptional man, an exceptional prophet, who has exceptional knowledge and who is seeking the highest knowledge. This search for knowledge is a journey into the central sphere, the unlimited sphere of knowledge, which requires guidance.

Before we go any further, I have to emphasize that the purpose of reading the Quranic text is to take knowledge from its pure source and quench our thirst for it. Would a thirsty person throw water on the floor, or would he rather drink it so that its essence runs through his body?

What good is knowledge if the person receiving it does not learn from it? Learning necessarily requires the person to apply that knowledge to himself. This is why I insist that the purpose of studying a story with an exceptional protagonist is not to assess Moses, but to infer and understand lessons that benefit each one of us.

If we were to replace Moses in this story with a normal person, it would not have the same effect on us, because we could be prey to arrogance and declare that we would not fall for the mistakes of such a normal person. Because the story happened to Moses, it is an indication that the mistakes made in it are serious, which in turn emphasizes the importance of the lessons learnt. May Allāh reward Moses with all the good for taking on the burden of the trials he faced in our place. Indeed, through this account, Allāh has taught us knowledge that leads us from darkness unto light.

9. The Young Man (*al-Fatā*)

And when Moses said unto his young man (al-fatā)...
[al-Kahf, XVIII: 60]

It is a very normal reaction to wonder about the identity of the *fatā* accompanying Moses. I have read many sources in which commentators attempted to discover the identity of this *fatā*, but based their arguments on insufficient evidence. Some of these attempts are completely unjustified and regretful, such as one that can be found in a commentary on a French translation of the Quran (made by a Muslim who is not French), in which Moses and his *fatā* are none other than Gilgamesh and Enkidu!

The Quranic text is very precise, concise, and perfect. The fact that the identity of the young man is not clarified means that this is not an important detail, compared to the message conveyed in the text. His identity is not significant for his role in the story. Commentators and readers of the Quran should not stop at the role of the young man. Maybe the *fatā* was mentioned to emphasize that the story happened when Moses was aging. Also, the Arabic word *fatā* can mean servant, assistant, follower, or even a disciple. These meanings allow us to conclude that the story happened long after Moses left Egypt for the first time. Therefore, he must have received the revelation by that time. This is also affirmed by Hadith accounts.

I think that the *fatā* was mentioned to bring to light a very important element in the story: the mistake of forgetting what should not be forgotten.

It was already introduced at the beginning of the story of Moses and al-Khiḍr, in verse 57 of Surat al-Kahf:

> *And who does greater wrong than one who has been reminded of the signs of his Lord, then turns away from them and forgets that which his hands have sent forth? Surely We have placed coverings over their hearts, such that they understand it not, and in their ears a deafness....* [al-Kahf, XVIII: 57]

> *And when Moses said unto his fatā: I shall continue on till I reach the junction of the two seas....* [al-Kahf, XVIII: 60]

This is the place where Moses met al-Khiḍr. The junction of the two seas is not a superficial or secondary detail. It is a key. *...till I reach the junction of the two seas, even if I journey for a long time.* [al-Kahf, XVIII: 60]

10. The Fish (al-Ḥūt)

Then when they reached the junction of the two (seas), they forgot their fish, and it made its way to the sea, burrowing away.

[al-Kahf, XVIII: 61]

It is clear from the context that the fish was a sign. As far as Moses was concerned, the fish was a sign that indicated the place and time of the meeting with al-Khiḍr. It is also clear that it was the *fatā* who was responsible for the fish. We can conclude from these verses that the *fatā* was not aware of the real function of the fish. He might have thought it was food.

What is worthy of note and striking at the same time is that Moses forgot to pay attention to the fish, the only and final indication of his meeting with al-Khiḍr. The junction of the two seas was a necessary detail, but not sufficient in itself. In a very significant manner, 'forgetting' (*al-nisyān*) was mentioned in this story 7 times: 4 times explicitly and 3 times in an implicit manner.

Then when they had passed beyond, he said to his fatā, "Bring us our meal. We have certainly met with weariness on this journey of ours." [al-Kahf, XVIII: 62] Only at that moment, the *fatā* remembered the fish, which he, in his simplicity, had thought was merely food. Any reader who is not alert would think, like the *fatā*, that the fish was food, since Moses asked for it to eat. I do not think that Moses was so hungry as to eat the only indication he had of meeting al-Khiḍr.

It is also noteworthy (and a particular sign) that the Arabic word for fish used in the story is *ḥūt* and not *samak*. The word *ḥūt* generally means a large fish, and in particular the whale which swallowed Jonah. We can only imagine how large that fish had to be to contain a living person inside it.

The word *ḥūt* occurs five times in the Quran:
- 2 times in this story
- 2 times in the story of Jonah (cf. al-Ṣāffāt, XXXVII: 142; al-Qalam, LXVIII: 48)
- and once in the plural form in the story of 'The Fish of the Sabbath' (cf. al-A'rāf, VII: 163)

Was it not exceptional that Moses, the great prophet and sage, should travel for a long distance while carrying a fish as a sign of the meeting between him and al-Khiḍr? Was it necessary? Was it not possible to arrange for that meeting between Moses and al-Khiḍr in a less exceptional way? Was the fish not the most unusual component in the story and the element that is most exceptional?

When an *intentional disruption* occurs in a perfectly arranged framework, it must be an allusion.

The fish (*ḥūt*) and the ant are the only creatures mentioned in a famous Hadith on knowledge (on *The superiority of the learned over the devout worshipper*).[22] And in

22. *It was reported that Abū Umāma al-Bāhilī said: "Two people were mentioned to the Prophet, peace and blessings upon him, one was a devout worshipper, the other learned, then he said: 'The superiority of the learned over the devout worshipper is like my superiority over the most inferior amongst you (in good deeds).' Then the*

a similar Hadith, reported by Ibn ʿAbbās, the fish is the only creature mentioned.[23] The word 'ants' occurs only once in the Quran, in Surat al-Naml, at the beginning of the story of Solomon and the Queen of Sheba, which also revolves around inspired knowledge.

In terms of numerology according to *Ḥisāb al-Jumal,*[24] the number that corresponds to the word *ḥūt*[25] in Arabic is 23 × 18 = 414.

23 - 18 = 5

23 + 18 = 41

These are signs for the knowledgeable.

---

messenger of Allāh said: 'Allāh and His angels and the inhabitants of the heavens and the earth, even the ant in its nest and the fish in the sea pray for the person who teaches people goodness.'" (Sunan Tirmidhī: 2609)

23. *It was reported that Ibn 'Abbas said: "Everything asks for forgiveness in favor of him who teaches goodness, even the fish in the sea."* (Sunan al-Darāmī: 347)

24. [Editor's footnote]: See the definition of *Ḥisāb al-Jumal* on p. 137.

25. [Editor's footnote]: ح (8) + و (6) + ت (400) = 414

11. Forgetfulness

He said, "Didst thou see? When we took refuge at the rock, indeed I forgot the fish - and naught made me neglect to mention it, save Satan." [al-Kahf, XVIII: 63]

...and it made its way to the sea [16th occurrence of the word sea][26] *burrowing away (saraban).* [al-Kahf, XVIII: 61]

...and it made its way to the sea [17th occurrence of the word sea][27] *in a wondrous manner. ('ajaban)!* [al-Kahf, XVIII: 63]

In terms of numerology according to *Ḥisāb al-Jumal,* the words *saraba(n)* and *'ajaba(n)*[28] = 263 - 76 = 187 = 17 × 11. The contents[29] of (33) = 17 × 33.

From the *fatā*'s words we understand that the point of meeting at the junction of the two seas, where the unusual event involving the fish occurred, was the rock. In numerology, the rock (*ṣakhrah*) = 926 = I remembered it (*'adhkurahu*). [30]

Even though what occurred was unusual, the *fatā* did not inform Moses of it. Maybe it happened while Moses was absent, even for only a moment, or while he was asleep and the *fatā* did not want to disturb him, then forgot to tell him. He forgot an exceptional thing. Satan caused him to forget. Do I have to emphasize the importance of this point and explain it?

The peril of knowledge is forgetfulness.

The *fatā*, after the above-cited verses, is no longer mentioned. In the following two verses we know that he is still present, but only through the verbs that refer to the third person dual (Moses and the *fatā*). In these two verses, we read that they retraced their steps, and that they actually found al-Khiḍr. At this point, the *fatā* vanishes from the text. His role ends and we find no mention of him after this.

This clearly indicates that the role of the *fatā* is to highlight the idea of forgetfulness and its reality, through the only sentence he uttered.

"indeed I forgot the fish - and naught made me neglect to mention it, save Satan and it made its way to the sea in a wondrous manner!" [al-Kahf, XVIII: 63]

He said, "That is what we were seeking!" So they turned back, retracing their steps. [al-Kahf, XVIII: 64]

---

26. [Editor's footnote]: The word "sea" (*baḥr*) is mentioned in the singular form in the Quran 33 times. See p. 32.

27. [Editor's footnote]: *Ibid.*

28. [Editor's footnote]: سَرَبًا : س (60) + ر (200) + ب (2) + ا (1) = 263
عَجَبًا : ع (70) + ج (3) + ب (2) + ا (1) = 76
In both cases, the *tanwīn* on the final letter is not taken into account in the numerology.

29. [Editor's footnote]: The contents of a number are the sum of all the whole numbers from 1 up to the given number. For instance, the contents of the number 5 are: (5) = 1 + 2 + 3 + 4 + 5 = 15.

30. [Editor's footnote]: *ṣakhrah,* الصخرة : ا (1) + ل (30) + ص (90) + خ (600) + ر (200) + ه (5) = 926
'adhkurahu, أَذْكُرَهُ : ا (1) + ذ (700) + ک (20) + ر (200) + ه (5) = 926

12. The Wise One

There they found a servant from among Our servants whom We had granted a mercy from Us and whom We had taught knowledge from Our Presence. [al-Kahf, XVIII: 65]

Is it possible for a knowledgeable person, who possesses inspired knowledge, to be anything but a servant of Allāh? All that person's states, in motion and in stillness, are founded upon worship, that is, absolute obedience, conforming to the Will of the One, Who created the heavens and the earth and their laws. By his absolute servitude, and by cleansing himself of all traces of his own self: *"And I did not do this upon my own command."* [al-Kahf, XVIII: 82], that person goes beyond the capabilities of the ego and its veils to integrate the laws of a universal system, in complete harmony. He becomes part of that system and thus becomes worthy of realizing the manifestations of the Truth. This is only possible through a mercy bestowed by the Merciful and a knowledge which He alone grants.

Moses said unto him, "Shall I follow thee, that thou mightest teach me some of that which thou hast been taught of sound judgment?" [al-Kahf, XVIII: 66] Through Moses's request, we clearly see his selflessness, even though he was the great prophet and had the highest levels of the knowledge of his time. This is why this story is an example of an extreme case through which we will have an opportunity to see Moses's reactions.

So what can we say about the reactions of ordinary people?

13. Patience

Our Master al-Khiḍr replied in a strikingly direct manner: *He said, "Truly thou wilt not be able to bear patiently with me."* [al-Kahf, XVIII: 67]. This is one of the most eye-catching and peculiar sentences in the Quran, all the more so because it occurs in this story and is emphasized 3 times, as in verses 67, 72 and 75, and 2 more times (with slightly different wording), as in verses 78 and 82. This sentence wonderfully seals the end of the story.

The effect of this sentence is increased even more by the fact that it is repeated in the same monotonic way, which strongly contrasts with the other verses of the story, which shift and bear a clear touch of suspense. The abstract idea included in this verse clearly contrasts with the other verses, which express physical and visual actions and events. The most important word in the sentence is *patiently*, which comes at the end of each of these verses. The story of Moses and al-Khiḍr ends with the word *patiently*. The word *patiently* occurs five times in this story.

Patience is the main virtue required of the five 'resolute prophets,' Noah, Abraham, Moses, Jesus Son of Mary, and Muḥammad, peace and blessings be upon them all.

He said, "Truly thou wilt not be able to bear patiently with me." [al-Kahf, XVIII: 67] Our Master al-Khiḍr justified and explained this, saying: *"And how canst thou bear patiently that which thou dost not encompass in awareness?"* [al-Kahf, XVIII: 68] In fact, this verse summarizes the main idea of the message conveyed in this story. This sentence is so concise, clear and powerful that it does not require any kind of comment on our part. *He said, "Thou wilt find me patient, if Allāh wills, and I shall not disobey thee in any matter."* [al-Kahf, XVIII: 69]

14. Silence, Gesture and Allusion

He said, "If thou wouldst follow me, then question me not about anything, until I make mention of it to thee." [al-Kahf, XVIII: 70] It is very important to examine this verse through a general and comprehensive understanding of the Quran.

After less than two or three pages of the story of our Master Moses and al-Khiḍr, at the beginning of the following surah, Surat Maryam, first Zechariah, then our Lady Mary were asked to remain silent before a miracle. They were only allowed to use gestures or allusions to communicate. **Because ordinary words are not adequate for extraordinary events.**

In Surat Maryam (XIX), we read: (10) *He said, "My Lord! Appoint for me a sign." He said, "Thy sign shall be that thou shalt not speak with men for three nights, [while thou art] sound."* (11) *So he came forth from the sanctuary unto his people, and signaled to them that they should glorify morning and evening."*

And in Surat Āl ʿImrān (III), verse 41, the word "gesture" (*ramz*) is used explicitly: *He said, "My Lord, appoint for me a sign." He said, "Your sign is that you shall not speak to the people for three days, save through gestures..."*

Why? What is the wisdom behind this?

Again, our Lady Mary is asked the same thing in verse 26 of Surat Maryam (XIX): ... *And if thou seest any human being, say, "Verily I have vowed a fast unto the Compassionate, so I shall not speak this day to any man."*

And when her people asked her about her son (verse 29): *Then she gestured to him. They said, "How shall we speak to one who is yet a child in the cradle?"* It is striking that the story of our Lady Mary in this surah begins with: *And remember (wa-dhkur).*

He said, "If thou wouldst follow me, then question me not about anything, till I make mention of it to thee." [al-Kahf, XVIII: 70] In this request, our Master al-Khiḍr informed Moses, peace be upon him, that his message had two parts:

- Elements of knowledge that are presented in a visual manner
- Oral commentary.

These two parts are comprehensive, complementary, and do not interfere with each other. The first is a visual expression and is unlimited, while the second is a limited, oral expression and is an aspect of the first. Each of the two parts is presented in a comprehensive manner.

Therefore, any question (from Moses) would corrupt the structure of the message. And any answer (from al-Khiḍr) in the form of an oral commentary would be invalid as long as the visual presentation has not completed itself. Any answer to a question would corrupt the structure of the message and create new situations that were not initially part of the structure of the story. In this case, any answer would descend to the level of the unprepared questioner. In reality, the questioner needs to rise up. To be able to rise up, he needs to rid himself of the heaviness of the material world.

Our Master al-Khiḍr must have been informed of our Master Moses's future reactions. This is why he confidently told Moses that he would be unable to have forbearance with him.

Nevertheless, even after that warning, what happened was meant to render the 'unseen' 'seen'. In Surat al-Kahf, we read: (103) *Say, "Shall I inform you who are the greatest losers in respect to their deeds?* (104) *Those whose efforts go astray in the life of this world, while they reckon that they are virtuous in their works.* (105) *They are those who disbelieve in the signs of their Lord, and in the meeting with Him. Thus their deeds have come to naught, and on the Day of Resurrection We shall assign them no weight."* [al-Kahf, XVIII: 103-105]

How tragic is the state of the people mentioned in verse 104!

In Surat al-'Ankabūt (verse 2): *Does mankind suppose that they will be left to say, "We believe," and that they will not be tried?* This is mercy! This shows people their reality. The trials show them their place on the path leading them from darkness to Light. They show them their weakness and their subtle faults, which are hidden within their sins. They show them their mistakes so that they can free themselves of them before it is too late, before the chance that they are given in this life ends and everything becomes fixed.

15. So They Went On

So they went on till.... [al-Kahf, XVIII: 71]

Our Master al-Khiḍr did not stop to give an oral sermon, filled with arrogance and pride in the knowledge he had mastered. It was not his intention to engage in debates. He did not utter one word. Instead, he gave our Master Moses a chance to reflect and meditate about what happened and what was said.

> *So they went on till, when they had embarked upon a ship, he made a hole therein. He said, "Didst thou make a hole in it in order to drown its people? Thou hast done a monstrous thing!"* [al-Kahf, XVIII: 71]

Our Master Moses forgot everything about his experience and wisdom. He forgot his promise and was not patient. In a hasty and impulsive manner, he objected. He objected and passed absolute judgment: "*Didst thou make a hole in it in order to drown its people?*" Then he judged his master's action: "*Thou hast done a monstrous thing!*"

Our Master Moses based his conclusions and his judgment on what is considered good sense and the limited manner of thinking that is commonly accepted. He saw what happened through a limited set of logical, hastily thought-out possibilities and explanations. He based his judgment on the assumption that his was the only valid reasoning and that it was suitable for any situation, as if the thought that occurred to his mind so hastily was enough for his judgment. The thoughts that came to his mind so hastily are thoughts that ordinary people use in their daily lives. Our Master Moses did not use his exceptional knowledge.

> (72) *He said, "Did I not say unto thee that thou wouldst not be able to bear patiently with me?"* (73) *He said, "Take me not to task for having forgotten, nor make me suffer much hardship on account of what I have done."* (74) *So they went on till....* [al-Kahf, XVIII: 72-74]

They must have walked for a long time. It was another chance for our Master Moses to reflect upon what had happened and what was said: *So they went on till they met a young boy, and he slew him. He said, "Didst thou slay a pure soul who had slain no other soul?"...* [al-Kahf, XVIII: 74] What evidence was there to prove it was *a pure soul*? And what evidence was there to prove that this person was killed (by al-Khiḍr) for any reason other than slaying another soul?

> (74) *"...a pure soul who had slain no other soul? Thou hast certainly done a terrible thing!"* (75) *He said, "Did I not say unto thee that thou wouldst not be able to bear patiently with me?"* (76) *He said, "If I question thee concerning aught after this, then keep my company no more. Thou hast attained sufficient excuse from me."* [al-Kahf, XVIII: 74-76]

In Ṣaḥīḥ Muslim, we read the saying of Prophet Muḥammad ﷺ: "*May Allāh have mercy on Moses; I wish he had been more patient so we could have learnt about what would have happened.*" And in another version, in the same chapter of Ṣaḥīḥ Muslim, we read: "*May the Mercy of Allāh be upon us and Moses, had he (Moses) shown patience, he would have seen wonders. But because he was afraid of being blamed by his companion, he said: 'If I ask you anything after this, keep not company with me. You will then have a valid excuse in my case.' Had he shown patience, he would have seen wonders.*" Since the Prophet ﷺ uttered these words, scholars of Islam have been wondering what could have happened if Moses had not apologized so adamantly. In fact, our Master Moses closed the door on his journey when he imposed an unnecessary and exaggerated limit on himself.

In this allusion, there is a lesson for the servants who burden themselves with unnecessary limits, when they could spare themselves. That is because these limits, namely exaggerated entreaties to the Merciful, exceed their capabilities.

So they went on till... [al-Kahf, XVIII: 77]. For the third time, our Master Moses was given a chance to reflect and meditate upon what had happened and what had been said:

> *So they went on till they came upon the people of a town and sought food from them. But they refused to show them any hospitality. Then they found therein a wall that was about to fall down; so he set it up straight. He said, "Hadst thou willed, thou couldst have taken a wage for it."* [al-Kahf, XVIII: 77]

For the third time, our Master Moses was not patient. He made the same mistakes, only this time he was the victim of the limit that he had set for himself. This is why our Master al-Khiḍr replied immediately and firmly:

> (78) *He said, "This is the parting between thee and me. I shall inform thee of the meaning of that which thou couldst not bear patiently:* (79) *As for the ship, it belonged to indigent people who worked the sea. I desired to damage it, for just beyond them was a king who was seizing every ship by force.* (80) *And as for the young boy, his parents were believers and we feared that he would make them suffer much through rebellion and disbelief.* (81) *So we desired that their Lord give them in exchange one who is better than him in purity, and nearer to mercy.* (82) *And as for the wall, it belonged to two orphan boys in the city, and beneath it was a treasure belonging to them. Their father was righteous, and thy Lord desired that they should reach their maturity and extract their treasure, as a mercy from thy Lord. And I did not do this upon my own command. This is the meaning of that which thou couldst not bear patiently."* [al-Kahf, XVIII: 78-82]

The last word in the story in Arabic is *patiently* (*ṣabrā(n)*). This root (*ṣabara*) is repeated throughout the account and in a unique way:

- 5 times as *ṣabrā(n)*, each time at the end of a verse
- And 2 times in different forms: *taṣbir* (you be patient) and *ṣābiran* (patient).

16. The Final Outcome

It is very interesting and highly significant that the story of our Master Moses and al-Khiḍr comes to an end without a clear exposition of its outcomes. It is noteworthy that the story comes to an end and another account is immediately started with no comment or brief summary of the lessons learnt. The lessons of the story were not expressed in words: this is an extremely important detail.

Expressing in words an unlimited, universal idea or concept is misleading. We mislead ourselves if we do so, because we confine unlimited thought to the limitations of words. We thus push that thought into the trap and vicissitudes of language and culture of a particular historical era.

Providing numerous examples without expressing the conclusions and lessons inferred from them, allows a more flexible, insightful, precise and better understanding. It is even more important because that approach is interactive and dynamic, as it obliges the mind to take a positive step toward understanding.

This is a justified style because it allows every reader the chance to comprehend according to his own personal capabilities and disposition. Therefore, this is a style that keeps the mind alert and frees it from the slumber of the negative and narrow nature of general, stereotypical thinking. This alertness and freedom are the main traits of the knowledge provided in the Quran; knowledge that is meant to change us in order to reach our true objectives in life. It is not a knowledge, the only purpose of which is to gain control over the material world, as if we were here to stay forever.

Allāh Most High said: (112) *He will say, "How many years did you tarry on earth?"* (113) *They will say, "We tarried a day or part of a day. But ask those who keep count."* (114) *He will say, "You tarried but a little, if you but knew."* (115) *"Did you suppose, then, that We created you frivolously, and that you would not be returned unto Us?"* [al-Mu'minūn, XXIII: 112-115]

And in Surat al-Rūm: *And on the Day when the Hour is come, the guilty will swear that they had tarried naught but an hour; thus were they perverted.* [al-Rūm, XXX: 55] Time is going increasingly faster, and worldly life is too short and precious to be wasted over blind, scattered and meaningless efforts.

Our Master Moses was, like all the other prophets, infallible in the message and the mission he was entrusted with. However, in other circumstances, where there were better or worse individuals, he was a human like all other fallible humans. Our Master Moses was a great prophet and he was sent like other prophets to show us the right path. This is precisely what he does in this story.

Prophets are human beings and well established models and examples for us. Every story in which they are portrayed is in fact an extreme example of a limited situation. In this story with the wise man, our Master Moses was chosen for the great knowledge he was famous for and the knowledge he learned during his stay at the Pharaonic court, which was the highest level of knowledge that the people of the earthly world possessed.

Nevertheless, Moses (peace be upon him), as a human being, made the most common and serious errors that anyone seeking knowledge, in particular the highest knowledge, could make:

- Even though he had been warned against impatience, he was impatient
- Even though he had been asked to show obedience and not to object, he forgot, broke his promise and was not obedient.

He would not have made these mistakes if he had renounced his ego and he would have seen the things that he saw for what they were. The manner in which his meeting with al-Khiḍr was arranged, the junction of the two seas, the fish, the *fatā* forgetting the fish, the only sentence that the *fatā* uttered... Every event that our Master Moses witnessed is worth meditating upon. Our Master Moses was invited to meditate upon what happened and what was said and to ask himself questions and wait to get the correct answers.

Our Master Moses, the human, was impatient and therefore broke his promise to obey and objected. Our Master al-Khiḍr wrecked the ship and disabled it to spare the poor sailors, who work hard at sea, the wrath of the angry king. This is how our Master al-Khiḍr began his visual presentation. Our Master Moses forgot how his own life had begun. He was thinking of the logical consequence of destroying the ship and forgot how his own life was spared in an illogical manner when he was still an infant, when his mother put him into the sea to protect him from an angry king. This is how he entered the king's palace and lived there in peace. When our Master al-Khiḍr killed the young boy, our Master Moses objected and strongly condemned what he considered a crime, committed against an innocent soul. Our Master Moses forgot all about the Egyptian he had killed, and who had not been guilty of killing another innocent soul, but was not an innocent soul. When our Master al-Khiḍr built the wall without taking any kind of reward for it, our Master Moses criticized him and said he could have been paid for it. Our Master Moses forgot how he had helped the two daughters of our Master Shuʿayb, the righteous, in the incident near the well in Midian (*Madyan* in Arabic). Our Master Moses did not take anything in exchange for his service, despite his state, which he described by saying: "*...My Lord! Truly I am in need of any good that Thou mayest send down upon me.*" [al-Qaṣaṣ, XXVIII: 24]

17. Knowledge

Our Master Moses did not benefit from his exceptional experience and did not link it to what he saw. He failed to see the relationship even though he was warned, and even though he should have been alert to grasp every sign he would encounter! What could we say about those who, out of ignorance, deny the relationship that exists between many points that are presented to them?

In principle, the elements of knowledge should be correct and valid like general information and data, but the fact that we acquire the data and information is not sufficient. People need to learn how to deal with the data and information and how to benefit from them. They need to learn how to rise to the level of the knowledge given and understand it correctly in order to be in harmony with it. That knowledge must become part of who they are and they must become part of that knowledge. This is only possible if seekers of knowledge are prepared in body, soul and spirit to receive it. This was the main aspect of the message of our Master al-Khiḍr in this story. The focus was not on the components or the substance of knowledge, but rather on the way and methodology of dealing with the data and information provided.

The Truth is one. Every single thing clearly points to this in one way or another. If people fail to see the Truth, it means that there are veils on their inner eye, not on the Truth Itself. In other words, the individual needs to change inwardly and outwardly to be qualified to receive knowledge. Knowledge must not be contracted, distorted or changed in order to lower it to the level of an unprepared person. Distortion and lowering characterize the style of teaching which aims to spread knowledge and simplify it for the common people.

Every subtle detail in the story of our Master Moses and al-Khiḍr can be understood on many levels, which are in harmony with and confirm one another; but this requires a great mastery of knowledge, caution and a meticulous approach to the text. In these 23 verses of Surat al-Kahf, which is surah number 18 in the Quran, we have gems of the highest knowledge. Understand the allusion.

In order to understand these gems, we should not use data and information that are confusing or useless, and put them between us and the Truth that we want to study; this would in fact veil the Truth from us. This is the essential mistake in the method adopted by common people in approaching data and information. This useless data and information make the distance between us and the gems of knowledge we seek to study wider, and the lights of knowledge are thus hidden from us. If we are to examine these gems of knowledge, we need to allow them to speak for themselves, independently of the din of useless data and information. In order for us to hear what these gems have to say, we need to listen in silence. In order to collect correct answers, we need to ask correct questions and wait for the answers from these pure gems coming from the Center and not from the periphery.

If we possess an answer to a question that we have, it does not mean that the answer we possess is the only answer there is. Look closely at what our Master al-Khiḍr said and did; anyone can infer that patience is commendable and that it is not right to judge things based on outward appearances. This is because there is a subtle wisdom in which Allāh Most High put one of His Secrets and which He taught to those upon whom He granted mercy and knowledge from His Presence.

Now that we have drawn upon this lesson that is known to everyone, is there anyone who can learn from it and apply it directly to the story of our Master Moses and al-Khiḍr!? Why was the meeting arranged to be at the junction of the two seas? Why was the fish used as a sign? Why were three examples used? Why were those examples in particular chosen and not others? Why were they presented in that order? What is that knowledge that exists beyond the two seas and a rock and a fish, and starts with a boat, then after an act of murder, ends with a hidden treasure? Two righteous parents and a young deceased boy, two young boys and a deceased righteous father...

I am not going to go into the details of these matters, and I will not suggest referring to the existing commentaries without reminding the reader that the most famous of these books were written in the Middle Ages. These commentaries represent roughly all trends of thought over thirteen centuries, from the Mediterranean to the Far East. However, I need to point out that we should not confuse a commentary which elucidates the meanings of the text with the act of using the text itself as an example to present a particular idea. For example, we can use the text of our Master Moses and al-Khiḍr as an example to illustrate the method that should be used to deal with a coded text and how to understand it.

As in the story, patience and memory are necessary tools in seeking knowledge and in approaching one of the texts that belongs to the knowledge of the elite. The hole in the ship represents in fact the defects or lacunae and deficient data and information that were deliberately put into these texts, so that they may be rejected, and therefore remain safe from the wrath of unjust 'kings.' This is particularly true in the case of knowledge aiming to serve the poor, but not unjust kings.

Similarly, the killing of the young boy is equivalent to the killing of new, young nihilist and unscrupulous philosophies and ideas, in order to replace them with others that are *better than them in purity, and nearer to mercy* (cf. al-Kahf, XVIII: 81) as a continuation of an ancient, valid philosophy.

Finally, the wall is like the sections in books which appear to offer explanation and elucidation without recompense, and at the same time offer consolidation that serves to strengthen existing ideas and concepts. But in fact the purpose is only to conceal a treasure and protect it, until it is excavated by its deserving, legitimate heirs.

Summary of Chapters

1. ***Wa idh* (and when):** Knowledge begins with remembering, remembering what Allāh Most High wants us to remember. The way to reach knowledge is through remembrance (of Allāh – *dhikr*); real remembrance is awareness.
2. **Knowledge:** Knowledge must be taken from its source for it to be pure and alive! The source of absolute, living knowledge is Allāh Most High, and only one who has had a chance to experience this can grasp the meaning of this statement.
3. **Adam and Knowledge:** None of the children of Adam, no matter how high their rank in the eyes of Allāh, can reach knowledge higher than the limits which have been defined for them. These limits are the reality of that person's vicegerency upon the earth. Vicegerency upon the earth requires knowledge. The children of Adam, the vicegerent on earth, have been allotted different ranks, and to each rank there is a corresponding amount of knowledge.
4. **Iblīs and ignorance:** The origin of all sources of ignorance. How can seekers claim to possess true knowledge if they do not keep away from and avoid the sources of ignorance?
5. **The errors of Iblīs:** It is only possible to begin to proceed on the path to Truth by avoiding mistakes; otherwise the seekers will go astray and be lost, like those *whose affair is ever [in] neglect.* (cf. al-Kahf, XVIII: 28)
6. **The human drama:** Between error and guidance, between corruption and righteousness, between the material and that which is not material, between the earthly world and the hereafter, between darkness and light.
7. **The escape:** This lies in the Quran and in reading it with consciousness, using the lessons therein and applying its knowledge in order to join the ranks of *those who hasten toward good deeds.* (cf. e.g. Āl 'Imrān, III: 114)
8. **Our Master Moses:** One of the greatest of the children of Adam, the vicegerent, who was offered great knowledge and spirit. He is a great model and example to follow.
9. ***Al-fatā:*** and the one sentence he uttered.
10. **The fish:** In sacred knowledge, a sign carries the most precious hints, which is why Satan works hard to veil it and deprive humans of it.
11. **Forgetting:** The peril of knowledge and the tool in Satan's hands.
12. **The Wise One:** One of the servants of Allāh; perfect servitude. He acknowledges Allāh's Divinity and is saved from the darkness and the veils of the divinization of his ego. This sets him free from its confusion to travel towards the Light.
13. **Patience:** It is a condition required of everyone who seeks knowledge. It is most manifest when the person controls the limitations of the ego, overcomes its rebellious nature, impedes its will, and does not trust its inclinations. In this way, patience that is accompanied by reliance on Allāh and trust in Him becomes a *beautiful patience.* (cf. e.g. Yusuf, XII: 83)
14. **Silence, gesture and allusion:** When we reach a level at which our ordinary and limited words fail to express higher knowledge, it becomes necessary to confine them to silence and replace them with the language of gesture and allusion.
15. **So they went on:** It is important for the individual to keep all that was learnt, with presence of mind, with every new step made on the path of knowledge.

16. **The final outcome:** It cannot be fragmented and distorted within the limitations of words.
17. **Knowledge:** True knowledge is not a pile of fragments that are, in the best case, like parts of dead bodies. True knowledge is in fact the means by which we deal with information and use it with a scale of priorities and a comprehensive vision, clearly differentiating between the means, objectives and goals.

Jalāl al-Dīn Rūmī met Shams Tabrīzī in Konya. When Tabrīzī asked him about the purpose of mathematics and the sciences, Jalāl al-Dīn answered: "*to discover the correct etiquette of the religion.*"

Shams al-Dīn responded calmly and confidently by saying: "*No, the purpose is to reach the Object of knowledge.*"

Then he recited the following poetic verse by Ḥakīm Sanā'ī Ghaznavī: "*If knowledge does not strip you of your ego, then ignorance is better.*"

Introduction to the Treatise

1. The Wafaq Square

The Wafaq square is called 'the magic square' in many languages. However, the word 'magic' is a very common word and is superficial; it is unpleasant-sounding to the ear, especially as it carries an undesirable connotation. In most languages the meaning of the word 'magic' has sunk low, and has become distasteful and provocative. On the other hand, *Wafaq* in Arabic is very meaningful and rich.[31] It was chosen by the masters and those who are knowledgeable in this science.

Wafaq, the noun, comes from the root *wafaqa*, the verb, which means to be accurate, matching, homogeneous, fitting, harmonious, etc. Anyone with good knowledge of the Arabic language is well aware of these meanings, which can fill at least two pages of an Arabic dictionary (cf. pages 262 and 263 of Ibn Manẓūr's *Lisān al-'Arab*; volume 12, Boulaq edition).

I shall leave the reader to take the time to delve into the various meanings and forms of the word *wafaqa*. He will find that all the forms are positive and their meanings convey a sense of joy and optimism. The derivatives of *wafaqa* give an impression of positivity and at the same time they are wonderfully suited to the meanings they express.

---

31. [Editor's footnote]: The word *wafaq* is used alone in Arabic. However, throughout the present translation, the term 'Wafaq Square' has been used.

2. The Triangle

The Arabic word *muthallath* describes anything composed of three elements and is used to designate the first two-dimensional geometric shape, the triangle.

It is also used to designate the 'Egyptian triangle,' which Europeans know as the Pythagorean theorem.

The same word refers also to the 'Triangle of the Gnostics,' which is called the *Tetraktys*.

Hermes (Hermes Trismegistus) is known in Arabic as *Hermes al muthallath bil ḥikmah*, which translates as 'Hermes the thrice wise.'

The Philosophers' Stone, or as it is called in Arabic 'The Stone of the Wise,' is also referred to as the 'Triangular Stone': body – soul – spirit.

All of these meanings are clearly and beautifully represented in the 3×3 Wafaq square.

3. The Treatise

The text that follows this introduction is written in the traditional style of a *treatise*. It is a style of writing that is intended for a limited circle of the elite in spiritual knowledge. This is why it is called a *treatise* (*risālah* (message) in Arabic) and not a book. A treatise is characterized by an intimate style, as if what is written is a secret conveyed to a trusted person, who is already familiar with the subject.

A treatise is written as a 'reminder' (*tadhkirah*) in the Sufi meaning of the word, and to record or emphasize some aspects of a particular matter. It is not the purpose of a treatise to teach in the general and ordinary sense of the word, nor to convince or to defend a particular idea or ideology. Some sections of a treatise may seem ordinary, trivial, superficial or of secondary importance. It is a very serious mistake to read these sections with proud neglect.

To start with, a treatise is not intended simply for anyone who can read a book! It is not one of those books that are completely open to everyone, which move sequentially, harmoniously and are accessible to anyone to read. The difference between those books and a treatise is that the clarity characterizing the treatise is also constant, but is appropriate to the level of the reader. Some sections can be read and understood by all types of readers, but others can only be read by those who are qualified to read them, and there is wisdom and mercy in this.

The text of the treatise is a series of remarks, notes and allusions. The sentences and words used are only keys to jog the memory. They are not written to express or reveal final conclusions. In the oral teaching tradition, they are simply pauses for explanation and commentary. They are also pauses for meditation and thought. Actually, a treatise should be read in the presence of a knowledgeable shaykh.

I have a friend, who is young and very intelligent and knowledgeable about general Islamic culture. He used to seek knowledge from a shaykh, who one day invited him to read a simple book on the Oneness of Allāh and Sufism. My friend did exactly that, and returned to his master, proud of having read the book in one night and of being ready to study and discuss it with him.

None of this happened. His shaykh simply asked him to re-read the book and come back again. My friend obeyed and the same thing happened again. His shaykh asked him: "Have you read the book?" He answered: "Certainly." My friend expected the shaykh to start a conversation about the book and its contents, but all the shaykh said each time was: "Go and read it again." This scenario was repeated eighteen times, and because my friend was well-mannered, he did not question, object or ask for explanations, and his trust in his shaykh was stronger than his own confusion. His shaykh instructed him to read the book time after time, until he felt his disciple started to have a general picture of the topic. This only happened after eighteen slow readings of the book. Only then did my friend read the book in the presence of his shaykh, when he almost knew it by heart.

During the discussion he had with his shaykh, he discovered that he had been arrogant and pretentious when he thought he had understood the book after the first reading. And even after eighteen readings, he only had a superficial idea about the general contents of the book. He discovered that he had read some phrases and even some entire sentences without paying attention to them. These were essential parts and needed long explanations. Only then did he realize that every word and every sentence in the text complements or explains the others. It is essential that every part of the text should be seen comprehensively and great attention should be paid to details. The details make the whole and the whole gives them meaning.

My friend understood that his approach to the book was hasty and lacked intelligence. He realized that his understanding was superficial and that his conclusions were limited and erroneous. He understood that his final conclusions and the true knowledge contained in the book cannot be found in the shells and dead bodies of words. They actually exist in the mind of the disciple, and the veils are lifted from this knowledge in stages.

4. Self-effacement

This experience was a harsh insult to my friend's self-esteem as a person who has a comprehensive mastery of modern culture and education.

Ibn al-'Arabī saw two doors: One was large and locked, and before it stood a crowd with so many people that it was difficult to approach it. The second one was open, no one was standing in front of it, but it was so narrow that one would need to bow one's head to pass through it. Ibn al-'Arabī knew that these were the two doors of attainment; the door of honor and the door of humility. I still remember how one of the followers of the Shaykh al-Akbar told me this story in the courtyard of the Umayyad Mosque in Damascus.

In fact, the door of honor was closed because none of the people crowded in front of it had real honor. There was no one at the open door of humility, because very few know what real humility is and to whom it is destined. Those who do know have already entered that door because they know that real humility is self-effacement.

One of the greatest veils that stands between seekers and the Truth and which blinds their inner eye is the veil of the self-centered ego, i.e. the 'I' which considers itself the center, with everything revolving around it.

My friend understood that the purpose was not to paste dead categories of knowledge upon a decadent mind, but to purify it deeply and awaken it.

One of my friends, a mathematics professor from the *Ecole Normale Supérieure* (College of Higher Education), told me how he started to think about the value of academic mathematics in the life of humans and how it contributed to their ascent towards the Truth. He was reflecting on that when, during a lecture, he witnessed one of his best professors conduct himself as one of the most lowly of the common people in the most deplorable state of rage.

He was prey to both his dead knowledge on the one hand, and to his ego on the other. His dead knowledge was a corpse without a soul. This situation is like a drowning person who, when calling for help, is thrown the handbook on swimming that he himself has written. This is why, at the higher level, it is not possible to separate knowledge from correct conduct.

5. The Point and the Circle

For people who read a treatise with a modern mentality, it may seem to them that it is a collection of disparate, incoherent items that are repetitive or contradictory. In reality, the subject of the treatise is presented in a comprehensive and universal way.

People of knowledge know that knowledge is one, like a point, and that a point is like a limitless circle. The *Gestalt* theory can help us understand this method.[32] Imagine a circle, two points and two small lines. If you were told that the point represents an eye, you might be doubtful, or even refuse to accept this. But if we were to arrange the 5 marks into a simple depiction of a face, each mark would have a meaning in relation to the whole, and would participate simultaneously in giving meaning to the other marks and to the whole, which in its turn gives the mark its meaning.

To understand the texts of the people of knowledge, we need to approach them in the manner of *Gestalt*. You may find in their writing that they say for example: "Look at your hand and how it has 5 fingers." It is necessary, although maybe annoyingly basic within the general context of the treatise, to consider the meaning of this sentence as a significant allusion and as a part of the whole. This is the first point.

Secondly, they want the disciples to reconsider what they claim to know, in order for them to see what the blind, slumbering and veiled eye of habit does not see. **That is because habit is a veil that covers the inner eye.** In fact, a sentence like: "Look at your hand" aims at directing meditation and thinking. This is what the people of knowledge call an *allusion*.

"Look at your hand and how it has 5 fingers." The key to this sentence is the word 'how'. The hand is a sea of knowledge. They do not want the disciple to be lost in it and drown. They want to direct his attention to focus in particular on one aspect of the symbolism of the five fingers, as an important reference point which they use and which enhances their topic of study.

Some cite other people's books and accomplishments. People of knowledge cite the accomplishments of the Creator.

---

32. Refer to the definition of the *Gestalt* theory on p. 136.

6. Allusion

Allusion is a very refined and lucid tool. It is essential to the method and style of traditional teaching on the path of Truth. This teaching focuses on the awakening and progress of the disciple, as well as on the knowledge he receives.

The knowledge that an allusion hints at can sometimes be expressed or presented in an almost or completely explicit manner. But it is considered an affront to knowledge if the reader is not prepared to receive it. That would be like sowing seeds onto a dead, untilled field or like generously offering something to hands that are clenched into fists. On the other hand, when a person is prepared and qualified to receive knowledge, an allusion remains at a higher level, unlike an explicit expression, which is like throwing a bundle out the window to anyone who might need it and might be glad to find it, or might simply put it on the side of the pavement.

Allusion is an open door. It respects the human being. It is an invitation to enter the principal gate of knowledge. Therefore, everyone takes what he is seeking every time he goes through the gate. Human beings and their states are constantly changing. Glorified be He Who changes hearts.

It is wise for a person to take what he is seeking at the proper moment. The most important characteristic of allusion is how it functions: it remains latent until **it is activated when a person is ready to receive it**. At the moment of comprehending an allusion, one feels a particular insight. It is like when one suddenly and clearly recalls something; it is as if the allusion is already present in the memory. Through this feeling, a person's memory comes back to life with a current that links it with the universal memory, in which here and there are one, and also, before and after are one.

This feeling is pure, true intuition and the direct connection to knowledge. It is part of 'unveiling' (*kashf*). It is completely different from the feeling that takes place with stubborn intellectual effort in which tension prevails. Tension implies a struggle between two contradicting forces. The result of this struggle is contraction (*qabḍ*). This is the complete opposite of the sought-after result, which is spiritual expansion (*basṭ*). Spiritual expansion allows a person to receive knowledge.

The disciple is taught that forcing understanding is pointless. It is as pointless as the case of someone who is trying to put the sea into a hole he has dug on the shore. He should simply wait for his efforts at self-improvement to bear fruit so that he can progress and rise to the required level of understanding. With self-effacement, the whole being of the disciple and all his powers work in the same direction - that of *basṭ* (expansion).

Unfortunately, very few pilgrims realize the meaning of *Ṭawāf*, circumambulating the Ka'bah.

The disciple is taught: "An answer to a question that you did not ask has no value or meaning in your eyes, as long as you did not ask the right question and as long as you are not worthy to receive and bear the answer that you are seeking."

True knowledge is made up of questions and answers that help us to change ourselves. It implies a transformation of our understanding, before and after. This is why we are in this world, the world of time and space.

❋❋❋❋

...Every person taking the first steps on the way towards central knowledge should have first meditated on and learnt from the stories of our Master Adam, Iblīs, the two sons of Adam, and Joseph and his brothers, as they are related in the Quran. Understanding, realizing and grasping the main idea in these stories are amongst the many indispensable steps on the path toward knowledge.

❋❋❋❋

What follows is a treatise entitled: *Treatise on the 3×3 Wafaq Square and its Relationship to the Canonical Prayer,* which is a vast subject.

The treatise is shortened in order to focus on a tradition concerned with sacred geometry, a science which has almost become extinct. It is also very closely linked to the science of numerology and the sciences of prayer. Each of these three aspects expresses the same one truth that phrases and words fall short of conveying.

Sacred geometry is a dense science that can be the foundation for theoretical and meditative research, as well as the basis for practical application. Practical application should not rely solely on a written source, as this is neither sufficient nor efficacious. It may also be very dangerous.

In fact, this treatise is a kind of introduction, demonstrating in a geometric and numerological manner the importance of the canonical prayer as one of the pillars of Islam. It shows how the canonical prayer was structured in an exact manner and how each one of the different postures and each detail pertaining to it conforms to specific foundations and rules that are in complete harmony with a universal system created by the Creator of the heavens and the earth, Exalted be He.

Treatise on the 3×3 Wafaq Square and Its Relationship to the Canonical Prayer

In the Name of Allāh, the Lord of Mercy, the Giver of Mercy

> (66) *Moses said unto him, "Shall I follow thee, that thou mightest teach me some of that which thou hast been taught of sound judgment?"* (67) *He said, "Truly thou wilt not be able to bear patiently with me.* (68) *And how canst thou bear patiently that which thou dost not encompass in awareness?"* (69) *He said, "Thou wilt find me patient, if Allāh wills, and I shall not disobey thee in any matter."* (70) *He said, "If thou wouldst follow me, then question me not about anything, till I make mention of it to thee."* (71) *So they went....* [al-Kahf, XVIII: 66-71]

An answer to a question that you did not ask has no value or meaning in your eyes, as long as you did not ask the right question and as long as you are not worthy to receive and bear the answer that you are seeking.

The purpose of the treatise is not to offer a scholarly explanation, but rather to draw attention to the great amount of information that some dense texts may contain, and to provide some indispensable data necessary for the correct approach to and precise study of the canonical prayer.

The 3×3 Wafaq Square

4	9	2
3	5	7
8	1	6

The 3×3 Wafaq square is the best and most perfect visual manifestation of the nine single-figure numbers and of the number 0, which is represented by the number 5, because arranged in this way they are as concentrated, interconnected and essentialized as they can be.

Let us think in a two-dimensional geometric manner and suppose that each of these nine numbers occupies the most compact and interconnected surface, which is the circle. Every number has the same importance as the other numbers, for we are looking at them here as meanings and symbols, not as representing a quantitative value. Now, if we bring together these 9 circles in the most compact and symmetrical way, we will get Fig. 1-B1.

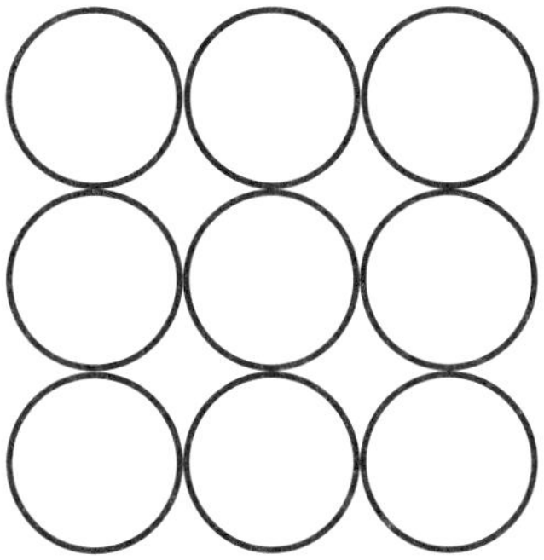

Fig. 1-B1

In between the 9 circles we can see empty spaces which separate them. Let us remove these spaces and imagine that we press our figure equally on each side, symmetrically, to make it more compact. What results is a square that is divided into 9 inner squares, each dedicated to one number (see Fig. 1-B).

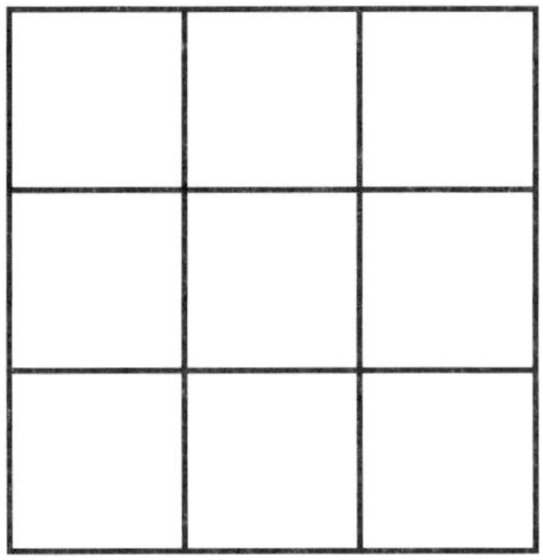

Fig. 1-B

If we distribute the numbers in the fields of this figure, based on the rules of the science of Wafaq squares, we will end up with only one possible distribution of the numbers. It is possible to look at this Wafaq square along eight different lines, as is the case with any 3×3 Wafaq square,[33] and the sum of the numbers along each line or side will be 15. The fact that there is no other distribution that achieves this result means that it is necessarily perfect. That is because unity and perfection complement each other (see Fig. 1).

4	9	2
3	5	7
8	1	6

Fig. 1

33. [Editor's Footnote]: These 8 lines are the 3 columns, 3 rows, and 2 diagonals in the square, each comprised of 3 boxes and their contents. 4 of these are sides, or *ribs* (in Arabic *ḍil‘*), while 4 pass through the central box containing 5.

The finest, simplest, and closest way to approach numbers in their visual form is in the 3×3 Wafaq square. As for the transcendent and unlimited knowledge that exists in the 3×3 Wafaq square, it is astounding.

Let us imagine the nine numbers, which, along with 0 constitute the 'alphabet' of the science of numerology. **5 is the central number,** linking the world of (1, 2, 3, 4) and the world of (6, 7, 8, 9), as shown in the following figure:

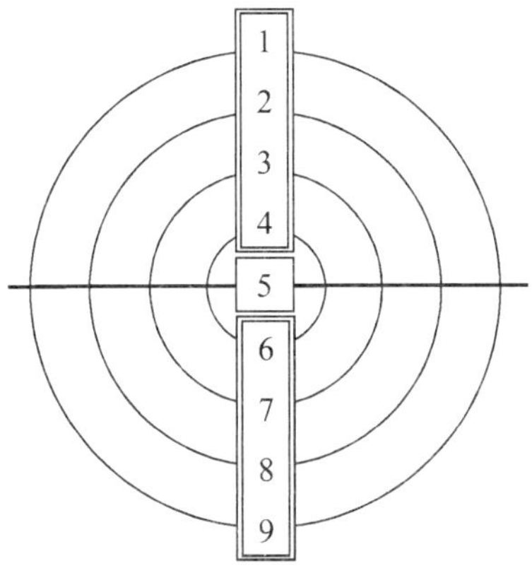

Imagine a pyramid, the four corners of the base of which are represented by the four numbers of one of the two worlds that we saw above. If you extend the edges of this pyramid, they will meet at the number 5 and then make another unlimited pyramid. This 'ends' in the infinite at the other four points, which represent the four numbers of the other world. Look at the human being, look at yourself, you will understand. Imagine the edges of these two pyramids being extended infinitely and being parts of infinite circles.

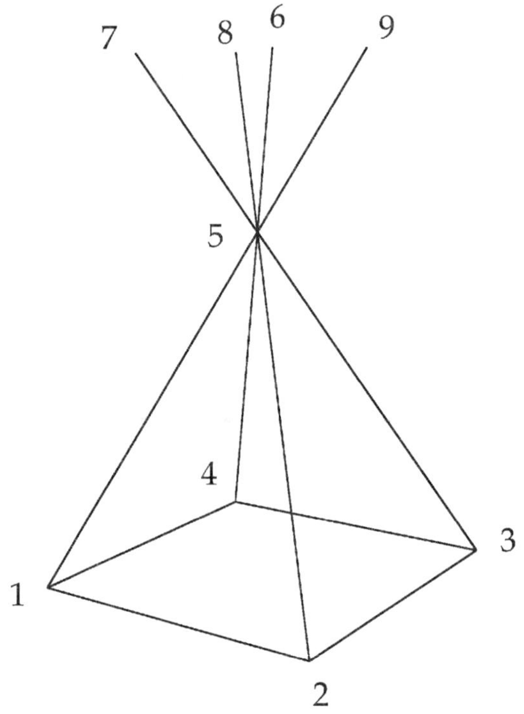

The human being occupies the highest rank amongst all creatures. Therefore he is the image of Knowledge. Did Allāh not create him in His Own Image?[34] The human being and the 3×3 Wafaq square are both representations of knowledge; we can therefore say that each represents the other.

It is said that 0, which is represented by 5, along with the 9 numbers, contain all knowledge. These numbers occupy the highest ranks of knowledge. They are also the sum of this knowledge and manifest themselves in our 3×3 Wafaq square.

34. It is reported that Abū Hurayrah (may Allāh be pleased with him) said: The Messenger of Allāh ﷺ said: "*When fighting, avoid the face, because Allāh created Adam in His Image.*" (Ṣaḥīḥ Muslim: 4731)

Indeed, the sum of the 9 numbers which represent the 3×3 Wafaq square and which, in turn, represents the human being, represented by Adam, is the contents of 9.

Contents of (9) = 1 + 2 + 3 + 4 + 5 + 6 + 7 + 8 + 9 = 45

Contents of (9) = $(9^2 + 9)/2 = 9 \times (9 + 1)/2 = 9 \times 10/2 = 9 \times 5 = 45$

According to the science of numerology, Adam (ادم) is portrayed as follows:

ا = 1

د = 4

م = 40

The sum of Adam is 45, which is the same as the sum of the 3×3 Wafaq square.

The sum along any side (rib, *ḍil'* in Arabic) of the 3×3 Wafaq square, whether vertical or horizontal, or even along the diagonals, is 15.

15 is simply the sum of the numbers along any of the sides (*ḍil'*) of the 3×3 Wafaq square.

When we understand that the 3×3 Wafaq square represents Adam, we will not be surprised to know that Eve (حواء) was created out of one rib[35] of Adam (45).

According to numerology Eve is portrayed as:

ح = 8

و = 6

ا = 1

حواء (Eve) = 15, and 15 is one of the ribs or 'sides' of Adam = 45. The letters of Eve are obviously contained in one of the sides of the 3×3 Wafaq square.

The people of knowledge were aware of this truth and one of them (al-Muttaqı al-Hindī) pointed it out briefly in one of his books, in less than one line.

15 = the central 5 × √9

45 = the central 5 × the final 9.

The name Adam is mentioned in the Quran 25 times:

5^2 = 25 times: the name 'Adam'.

4^2 = 16 times: referring to Adam himself.

3^2 = 9 times: referring to the children or progeny of Adam.

Look at your hands as if for the first time. They carry the history of your life and an interconnected image of your body and your body parts. Your hands represent all of you, just as you represent humanity. They contain universal proportions and

35. [Editor's footnote]: In Arabic, the edge of a geometric shape and any of the human ribs are both called *ḍil'* (ضلع). So the image is very clear in Arabic!

values. They contain the Golden Ratio which is embodied by the regular 5-sided figure, the pentagon.[36]

Your two hands are at once a diagram and an image of the 28 letters of the Arabic alphabet, and of the 28 lunar mansions.[37]

Numerically, the two hands symbolize the 28 cards of the Tarot, whereas the two feet symbolize the other 28 cards and the head the 22 cards of the Major Arcana. Is it not enough to look at the bones?

Therefore, we have 56 cards for 4 limbs (hands and feet) and 22 cards for the fifth (the head): 4 + 1 = 5 and 56 + 22 = 78. [38]

78 cards correspond to the 78 isolated letters that open some surahs of the Quran.[39] These 78 isolated letters consist of 14 distinct letters, which is half the Arabic alphabet. The 14 letters are contained in the right hand, which represents the four eternal holy places. The fifth[40] is the one place, which became sanctified by the existence of the Perfect Man (*al-insān al-kāmil*).

The right hand symbolizes the Triangle of the Gnostics mentioned above. The left hand completes this symbolism, by comparison and contrast.

The hands represent a science. Do you also realize that the Arabic word for 'hand' (*yad*) numerologically[41] equals 14? 4 and 10 as in the Tarot.[42] 4 and 10 as in the Triangle of the Gnostics.[43]

Do you need to go over this again? Look at yourself. What is the first thing that you can best see and meditate upon? It is yourself, the intermediator represented by your body. Every part of your body is symbolic and representative and can be a point of reference. Beware of misbehaving and misusing this body or going astray; Adam misbehaved whilst in Paradise!

Your thumb is in proportion to your hand, and in the same way your hand is in proportion to your body, and your body is in proportion to its surroundings and

---

36. [Editor's footnote] The diagonals of a regular pentagon are in a Golden Ratio with its sides, and the subdivisions of the pentagram formed by these diagonals give further lengths in the Golden Ratio.

37. [Editor's footnote] 14 is the number of phalanges in each hand – each of the four fingers has three phalanges, but the thumb has two - and the total for both hands is therefore 28. The same applies to the toes and the feet.

38. [Editor's footnote] The four suits of the Minor Arcana have 14 cards each = 56 cards, while the Major Arcana has 22 cards.

39. [Editor's footnote] The isolated letters referred to here are those that appear at the beginning of certain surahs, such as *Alif, Lām, Mīm* at the beginning of Surat al-Baqarah.

40. [Editor's footnote]: These five holy places, which are alluded to in Surat al-Tīn, are: [1] *Shām* (the Levant), [2] *Quds* (Jerusalem), [3] Mecca, [4] Mont Sinai and [5] Medina (sanctified by the presence of the Prophet Muḥammad).

41. [Editor's footnote]: ی (10) + د (4) = 14.

42. [Editor's footnote] Each of the four suits of the Minor Arcana has 14 cards, consisting of 4 'court' cards and 10 'pip' cards.

43. See the definition of the Tetraktys or the Triangle of the Gnostics on p. 134. [Editor's comment]: The Tetraktys consists of 10 numbers arranged in 4 rows.

our surroundings are in proportion to the earth and the earth is in proportion to the universe...

In the universal framework, measurements and proportions are essential, and they are certainly so by the wisdom of Allāh.

❋❋❋❋

The Holy Quran is part of a universal order based on an intentional order that is so miraculous and so meaningful! Its verses were received and gathered according to the chronological order of revelation. Later, each word in the Quran took its rightful place and spelling according to the instructions that the Prophet ﷺ received from Allāh (glorified and exalted be He) through our Master Gabriel (may Allāh be pleased with him) such that he placed the verses into a certain order and made it into a universal message that is in perfect equilibrium and which has an outward and an inward expression.

Numerology is one of the main keys to an accurate understanding of this universal order.

When our Master Gabriel (may Allāh be pleased with him) first appeared to the Prophet saying three times: *"Recite"* (*iqra'*, literally read), the unscriptured (*'ummiy*)[44] Prophet ﷺ was seized by the highest degree of fear he had ever experienced in his life. Each time, confronted with the perplexing command: *"Recite,"* he answered: "I am unable to read!" until our Master Gabriel said to the Prophet ﷺ: *"Recite in the Name of thy Lord...."* [al-'Alaq, XCVI: 1] At the second word of the revelation, *in the Name*, our Master Gabriel's demand began to become clear, and then it was completely clear with the third word *thy Lord* (*rabbika*, ربك).

ربك = (200 + 020 + 002). It is possible to give an extensive explanation of this, using the science of numerology.[45]

The *basmalah* (the phrase *Bismillāhi r-raḥmāni r-raḥīm*) is the first verse in the order of the Quran. It is the beginning of the Fātiḥah. The Fātiḥah is the first surah in the order of the surahs of the Quran. It was the fifth to be revealed.

The first complete surah in the ordering of the Quran is the fifth in order of revelation (al-Fātiḥah). The last complete surah in the revelation of the Quran is the fifth from the end (Surat al-Naṣr).[46]

In a very essentialized and condensed explanation of the meaning of the letter *ba'*, with which the *basmalah* begins, we can say that Allāh (glorified and exalted be He) is saying: "By Me, ever was what was, and by Me, ever shall be what shall be."

44. [Editor's footnote]: The Arabic word *'ummiy* is translated as 'unscriptured' due to the fact that the Prophet ﷺ had no knowledge of the earlier revealed scriptures. Refer to the chapter *al-'ummiyūn* in *Kitāb al-Iḥsān* by Suleiman Joukhadar, Dār al-Fikr, Damascus, 2016; the English translation is forthcoming.

45. [Editor's footnote]: ر (200) + ب (2) + ک (20).

46. It is reported that 'Ubaidullāh b. Abdillāh b. 'Utbah said: *Ibn 'Abbas told me: "Do you know that the last surah of the Quran to be revealed was revealed in its entirety?" I said: "Yes, 'When Allāh's Help and Victory come'* (Surat al-Naṣr)*." He said: "You have spoken the truth."*

All the 114 surahs of the Quran begin with the *basmalah* except for Surat al-Tawbah, which is remarkable (114 – 1 = 113). This is the only surah where the first verse begins with the letter *bā'*: (*barā'ah*, براءة); therefore all the surahs of the Quran begin with the letter *bā'*.

The change is intentional, and is compensated for in the overall order; this is indeed a sign to pay attention to.

If we look for the *basmalah* that is missing at the beginning of Surat al-Tawbah, we will find it in Surat al-Naml. This is the only *basmalah* that occurs in the text of a surah, not at the beginning of it. This is a change that is compensated for and significant at the same time. In other words, the *basmalah* occurs (113 + 1 = 114) times, like the 114 surahs.

The foremost among the people of knowledge have pointed to the story of our Master Solomon related in Surat al-Naml, especially in some of its details, such as the letter which our Master Solomon sent to the Queen of Sheba and the details related to her throne. They point out verse 30 of this noble surah, which includes a *basmalah* within the Quranic text, which is not the *basmalah* at the beginning of that surah.

The letter *bā'* is the second letter of the Arabic alphabet and the first to appear in the greatest of alphabetical structures and the greatest written reference book, which is the Quran. The letter *bā'* is the second letter of the Arabic alphabet and the first of the first sentence in the Noble Quran: *Bismillāhi r-raḥmāni r-raḥīm.*

Through our Master Gabriel, Allāh (glorified and exalted be He) revealed the Quran to the unscriptured Prophet ﷺ over a period of 23 years, in the form of sequences of verses.

Muḥammad = 92 = 4 × 23 [47]

The first letter in the Quran after it was put into its present order is the letter *bā'*. Additionally, the first letter of the first word revealed is the letter *alif* (the first letter of the Arabic alphabet), where in Surat al-ʿAlaq, Allāh (glorified and exalted be He) says: *"Recite."*

Islam is based on 5 pillars,[48] meaning that Islam was founded upon the pentagon, which best embodies the Golden Ratio. Four of the pillars, in reverse order of importance, are: *ḥajj* (pilgrimage), *zakāh* (giving charity), *ṣawm* (fasting), *ṣalāh* (canonical prayer). These are indispensable, but they are useless without the fifth, which is the first in order of importance: the *shahādah* (uttering the declaration of faith: *lā ilāha illā Llāh Muḥammad rasūlu Llāh*). In other words, there are five (5) pillars in total. On the one hand, there is the pilgrimage (also founded on 5 pillars),[49] which is closely linked to

47. [Editor's footnote]: م (40) + ح (8) + م (40) + د (4) = 92 = 4 × 23

48. [Editor's footnote]: The Arabic word *rukn* (pl. *arkān*) translated here as 'pillar,' literally means 'corner,' as in the vertices of a polygon.

49. According to the Shāfiʿī school, there are five pillars of Ḥajj: 1. wearing the *Iḥrām*, 2. standing on the mountain of 'Arafāh, 3. performing *Ṭawāf al-Ifāḍah*, 4. striving between *Ṣafā* and *Marwah* and 5. shaving the head or shortening the hair.

Mecca and the Black Stone in one corner of the Ka'bah, the name of which is derived from *ka'b* and *muka'ab* (cube). On the other hand we have the *Shahādah.*

The acts of worship that were revealed to Prophet Muḥammad ﷺ through our Master Gabriel are based on precise numerical foundations; they are defined by numerical values. Studying one of these values suffices to understand the remaining pillars. The canonical prayer as we know it was revealed to Prophet Muḥammad ﷺ in his mysterious and well-known nocturnal journey that took him from the foot of the Ka'bah in Mecca, to Jerusalem, then into other dimensions.

The story of this journey, *al-Isrā' wal-Mi'rāj* is first mentioned at the beginning of Surat al-Isrā'. Then the story is completed after 35 surahs, in Surat al-Najm. al-Isrā' is surah number 17, and al-Najm is surah number 53. This is a great sign!

In this journey, the Prophet ﷺ reached places that even our Master Gabriel, peace be upon him, did not reach. During that night journey, the Prophet ﷺ received great revelations. Perhaps the greatest revelation he received with relation to the life of Muslims was the obligation of performing the canonical prayers and most particularly the number of these prayers, which are performed today facing the Ka'bah, exemplifying how the canonical prayer was revealed to the Prophet ﷺ in another world.

The canonical prayer is the divine message that the Prophet ﷺ received during the Night Journey. It is a direct message that Allāh revealed to the Prophet ﷺ when He said to him: "*O Muḥammad, there are five prayers every day and night and each prayer is equivalent to ten, which add up to fifty prayers.*"[50] All the other religious duties were transmitted through our Master Gabriel.

Surat al-Isrā' begins as follows: *Glory be to Him Who carried His servant by night from the Sacred Mosque to the Farthest Mosque, whose precincts We have blessed, that We might show him some of Our signs. Truly He is the Hearer, the Seer.* Prayer is one of these signs.

50. Ṣaḥīḥ Muslim 234

It is hard to conceive what an incredible, unlimited amount of information is contained in the canonical prayer. However, it is possible to catch a glimpse of this through two-dimensional sacred geometry, based on numerical foundations and a set of primary, basic, and grand rules, in complete harmony with the purpose and symbolism of the 3×3 Wafaq square (Fig. 1).

4	9	2
3	5	7
8	1	6

Fig. 1

4	9	2
3	5	7
8	1	6

Fig. 1-A

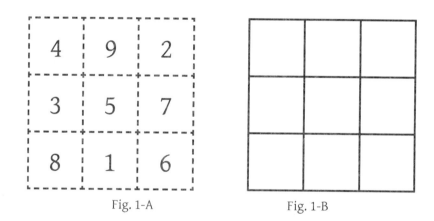

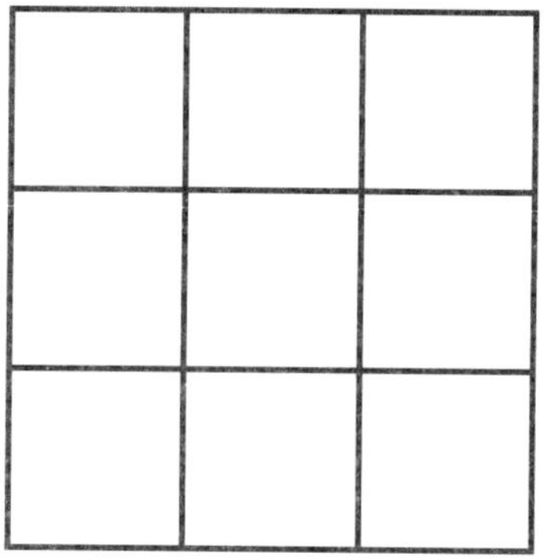

Fig. 1-B

Primary, Basic, and Grand Rules

In sacred geometry there are two distinct operations that are impossible to separate. The first operation defines the second: Fig 1-B and Fig 1-A. These are exquisite, pure examples of sacred geometry.

The figures are in complete harmony with the purpose and symbolism of the 3×3 Wafaq square. Therefore, they are constructed on the basis of a number of primary, basic, and grand rules, related to two-dimensional sacred geometry.

Primary Rules

Primary rules include:

- The use of no more than three tools: one to make a mark, a second to trace straight lines and a third to draw circles. It is to be noted that set squares and protractors, which are very common tools in this context, are inappropriate.
- Only the lines forming a final figure, which are lines and forms of defined shapes as in Fig. 1-B, have a definite beginning and a definite end.

Basic Rules

Among the basic rules used in the construction of Figures 1-B and 1-A, we can mention:

- Achieving the desired shape with absolute precision and the utmost simplicity with the minimum number of operations, such that no step can be dispensed with and such that there is no shorter, simpler or more precise way to achieve the shape.

This last rule compels us to count each step in any given geometric operation in order to determine which is the shortest, simplest and most precise way. This calculation provides us with definitive numbers, expressing the lines, circles and intersections. These numbers are the elements of the grand rules of sacred geometry.

Grand Rules

Numbers represent the highest level of knowledge and everything is organized and based on them. Therefore, numbers are the greatest and ultimate criterion in determining whether or not we are confronted with a work of sacred geometry, based on the properties of numbers and their meanings, according to the principles of equivalence, measure, comparison and balance and the other foundational principles related to the science of numbers.

According to the teachings I have received, no shape in sacred geometry can be invented by a human being. That shape or figure already exists because it is a manifestation of eternal Truth. Any figure of sacred geometry is in fact an unveiling and is given to human beings from the Higher Source. A figure conforming to the principles of sacred geometry can be used in the study of advanced, essential knowledge.

Only then will that figure unveil its mysteries, thanks to our intelligent and meticulous reading of its construction, step by step. Every movement, operation, step and even intersection: everything is recorded and every movement is related to the work. All these elements are an essential part of the comprehensive message. All these details must be studied in light of the interrelationship between sacred knowledge and numerology with the support of pure, authentic acts of worship revealed by Allāh Most High. Similarly, support must be sought by contemplating creation which is an aspect or a manifestation of one and the same Truth.

In the study of sacred knowledge the sources of pure, true revelation, such as the Quran and valid Hadith, as well as acts of worship (canonical prayer and pilgrimage, for instance) are references used to interpret and to be interpreted. The realm of creation, which is based on the same principles used in numerology, sacred science and revelation, helps to provide an understanding which is simpler for minds to grasp and a more tangible understanding of concepts that are extremely abstract and dense, such as numbers. Dealing with a given number requires an in-depth understanding of its meaning, what it hints at and the properties it contains.

This understanding, which is reached, developed and enhanced by grasping it over the years from the Divine Source of Knowledge, or through oral teaching, is only an introduction to and evidence of that Direct Source. It is impossible to acquire such an understanding, based only on written sources. It is also inaccurate and unsound to give a definition of a number in our ordinary words, just as it is inaccurate and of no value to put a limitation on the infinite. It is such a transgression, that it is as if we are asking that which is elevated to descend here below.

In order for us to hear the silent eloquence of a number, we must give it the attention it merits, and a silence filled with reverence.

The manifested world of creation is introductory and primary as well as a final means of understanding in our study of numbers. Revelation (Quran, Hadith, acts of worship) and sacred geometry are higher levels of understanding numbers. Numbers themselves are best placed to explain numbers through their inner relationships and because of their correspondences and similarities...

Steps for Forming a 3×3 Wafaq Square

Let us tie a rope into a loop with 3 + 4 + 5 = 12 knots.

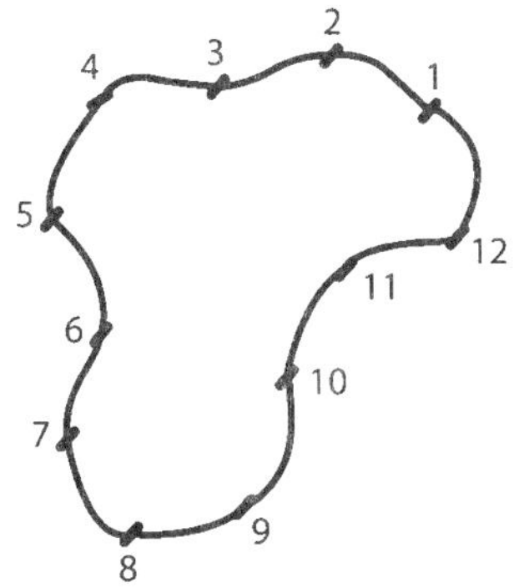

Step 1: Fix the rope into place at knot number 1.

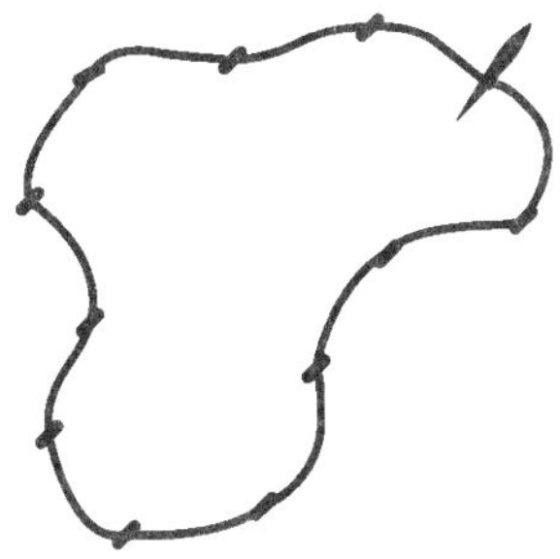

Step 2: Count three knots, pull the rope tight and fix it into place.

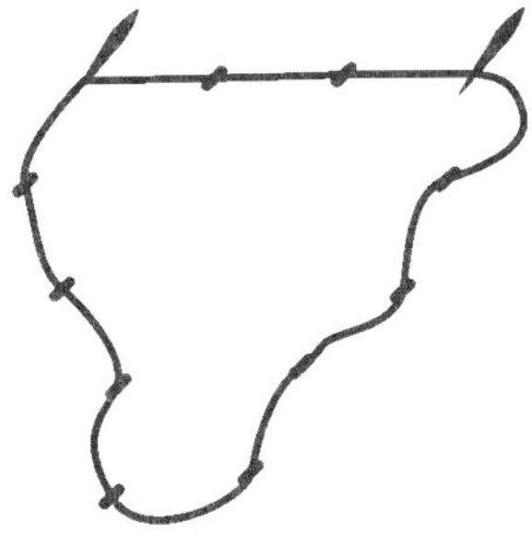

Step 3: Count 4 knots, pull the rope downwards, and you will have a 3, 4, 5 triangle; do not fix it into place at the 4th knot below this point.

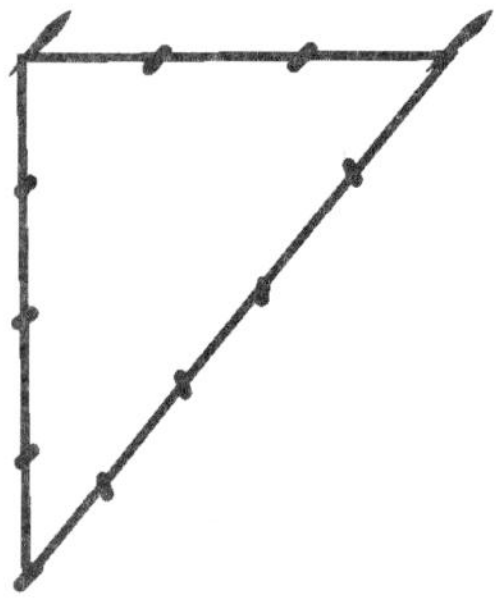

Step 4: Fix it into place at the 3rd knot below this point.

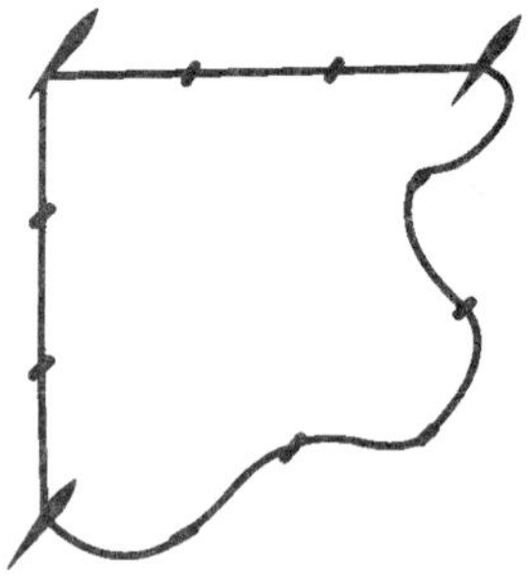

Step 5: Count 3 knots, pull the rope sideways, and tighten it into place.

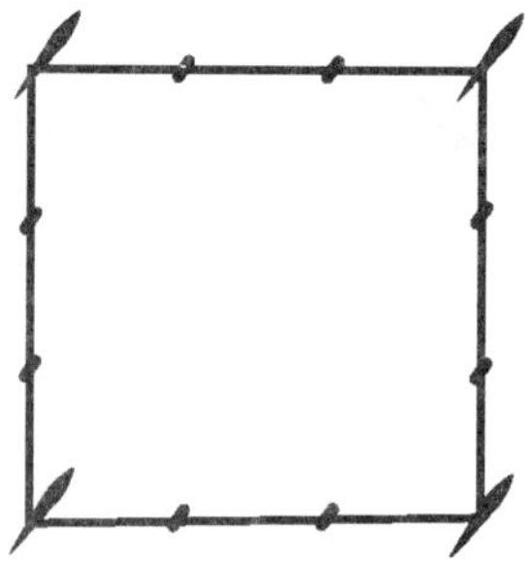

With these 5 preparatory steps we end up with the 12 points we need to form the 3×3 Wafaq square.

The 5 steps we took to reach Fig. 1-B are necessary, based on the 'basic rules' of extreme economy in the geometric construction of the 3×3 Wafaq square.

The fact that there are 5 steps puts them into complete harmony and coherence with the fundamental role of the number 5 in the 3×3 Wafaq square.

Before studying Fig. 1-B and the number 5 in detail, it is necessary to know the role of the operation described above in the comprehensive geometric construction of the 3×3 Wafaq square. In other words, we will apply the *Gestalt* theory in order to understand the inner relationship between the part and the whole.

These 5 preparatory steps made all the other steps possible to reach Fig. 1-B and Fig. 1.

The 3×3 Wafaq square is made of 8 linear elements and 9 figures (numbers or letters), adding up to 17 elements in total.

The presentation of these facts is a primary means of checking that the work corresponds to some of the most important fundamental principles of the universal system, before undertaking a profound study of Fig. 1. Such a study gives us an idea of a pure and a very lofty example of sacred geometry.

Fig. 1 was made through 5 preparatory steps.

Step 5 was of utmost importance, especially because it clearly corresponds to the central number 5 in our 3×3 Wafaq square. By Allāh's Mercy, the signs are given to guide us. Everything is inviting us to follow the number 5 and study it.

I am not going to pause to dwell on the well-known role of the number 5 in *phi* (the Golden Ratio), or on its other properties. I would like, however, to focus on a numerical approach that I have received by the Grace of Allāh. Back in 450 BC it was well known among Pythagoreans that the number 5 leads to 17, but this knowledge was lost and became a mystery. This is what I was told by a professor, who teaches sacred geometry in an international university, when I shared it with him.

Solving the mystery of the relation between 5 and 17 is a process of utmost importance and is extremely revealing. It is also possible to solve it in various other unconventional ways. We divide the number to be studied into its closest integer components, then we divide these components in the same way and carry on the same operation until each reaches 1.

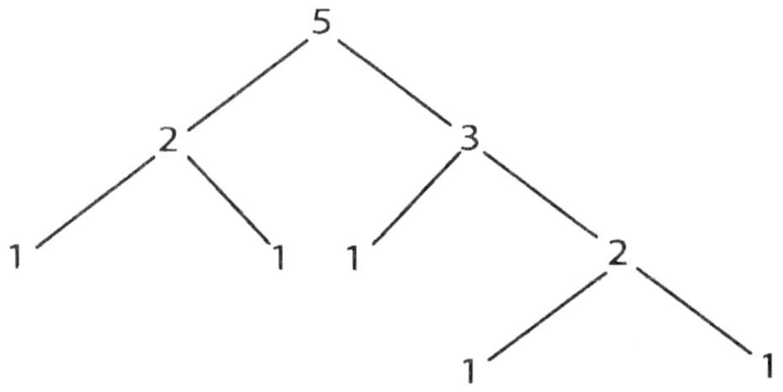

Fig. 5-A1

If we change the numbers in Fig. 5-A1 into points we will have Fig. 5-B, which is composed of 9 points and 8 connecting lines. In other words: 9 + 8 = 17 elements.

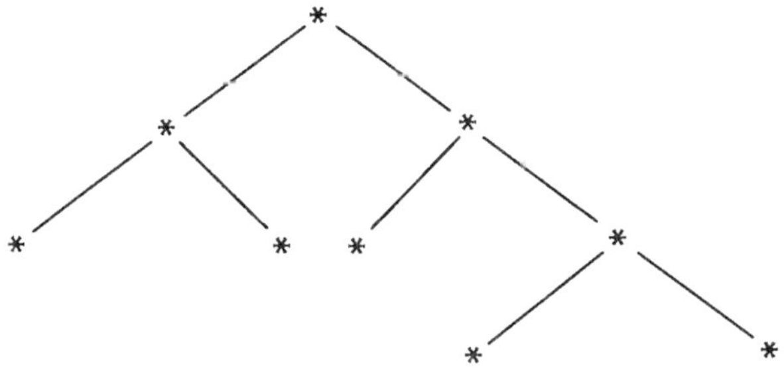

Fig. 5-B

Let us give each point in Fig. 5-B a number as in Fig. 5-C.

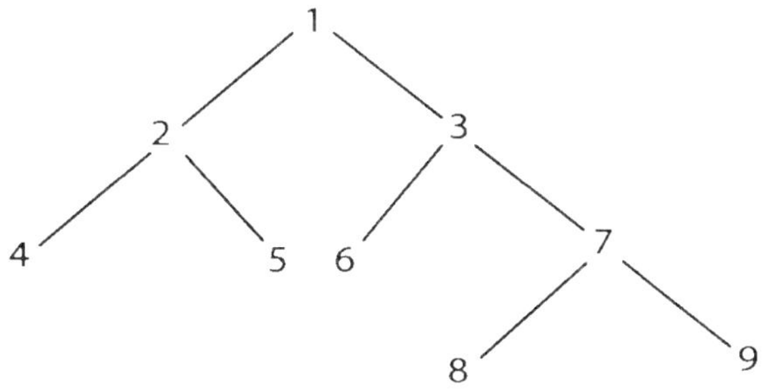

Fig. 5-C

The numbers used in Fig. 5-A1 are:

5, 3, 2, 2, 1, 1, 1, 1, 1 adding up to 17 as shown in Fig. 5-D. The individual numbers used in Fig. 5-A1 without repetition are: 1, 2, 3, 5. They add up to 11, as shown in Fig 5-E.

5		5 = 5
3		3 = 3
2		4 = 2 + 2
1		5 = 1 + 1 + 1 + 1 + 1
—		—
11	....	17

Fig. 5-E Fig. 5-D

The figures 5-A1, 5-B, 5-C, 5-D, 5-E were reached and derived through simple and clear logic, based on rules which form a concise and independent realm. This is also a manifestation of Truth, as is Fig. 1, which was achieved by a different simple logic, as well as primary, basic, and grand rules.

One of the correct acts of worship, received purely and directly from the Source of Revelation, confirms and explains our work and can be interpreted as follows:

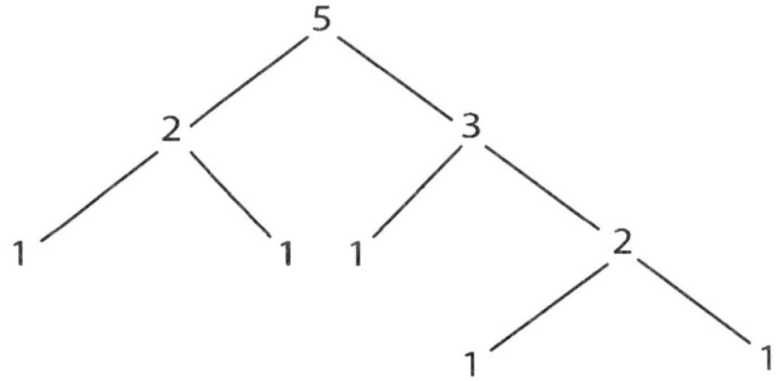

Fig. 5-A1

As in Fig. 5-A1, the canonical prayer, which is one of the 5 pillars of Islam, consists of:

- 5 prescribed daily prayers
  - 2 in the light of day
    - 1 prayer, *Ẓuhr*, which starts at midday
    - 1 prayer, *'Aṣr*, which ends at the end of the day
  - 3 prayers when the sun has set. These 3 prayers are 1 and 2:
    - 1 prayer, *'Ishā'* or 'night' prayer
    - 2 prayers: *Maghrib* and *Fajr*

The *'Ishā'*, *Ẓuhr* and *'Aṣr* prayers are all composed of four *raka'āt*, which explains why they figure on the same line in Fig 5-A1. However, *'Ishā'* is separate from *Ẓuhr* and *'Aṣr*, because it is performed at night and the night prayers are performed aloud.

The *Maghrib* and *Fajr* prayers have one thing in common. Therefore we can put them in the same group. The common factor is the time during which they are performed, which is determined by the sun. The time frame of the *Maghrib* prayer begins with sunset; and to complete the balance, the time frame of the Fajr prayer ends with sunrise. This is clear in Fig. 5-A2.

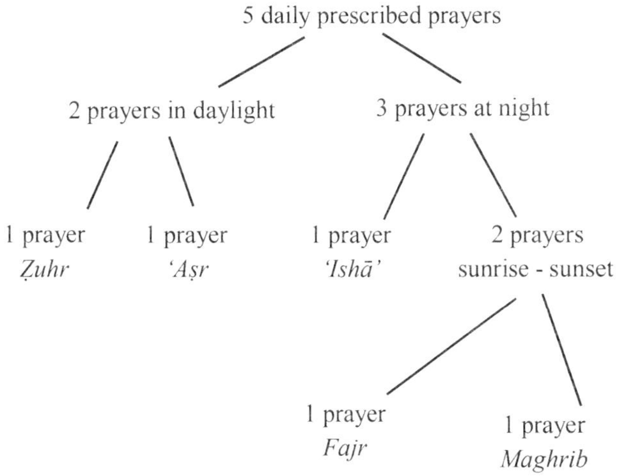

Fig. 5-A2

Up to this point, all that has been said may be no more than coincidence. However, a detailed study gives us absolute and more profound confirmation that the relationships and correspondences between sacred knowledge, the science of numbers and true acts of worship are relationships that are true and perfect, because each of the three is a manifestation of the same Truth.

In the 3×3 Wafaq square, there are 8 lines and 9 boxes for nine numbers as in Fig. 1-B. Similarly, Fig. 5-B is also made up of 9 points and 8 lines. If we continue and compare Figures 5-B and 5-C we will find:

- Point 1 gives us:
  - 1 set of daily canonical prayers.
- Points 2 and 3 give us:
  - 2 prescribed prayers in daylight – 3 prescribed prayers after sundown.
- Points 2 and 3 give us:
  - Each of the points 4, 5, 6 (remember the similarity between *Ẓuhr*, *'Aṣr*, *'Ishā'*) and point 7.
- Point 7 gives us:
  - Points 8 and 9, which are the last (remember the similarity between *Maghrib* and *Fajr*).

Circumambulation of the Ka'bah is another confirmation, as we can clearly see in the following figure:

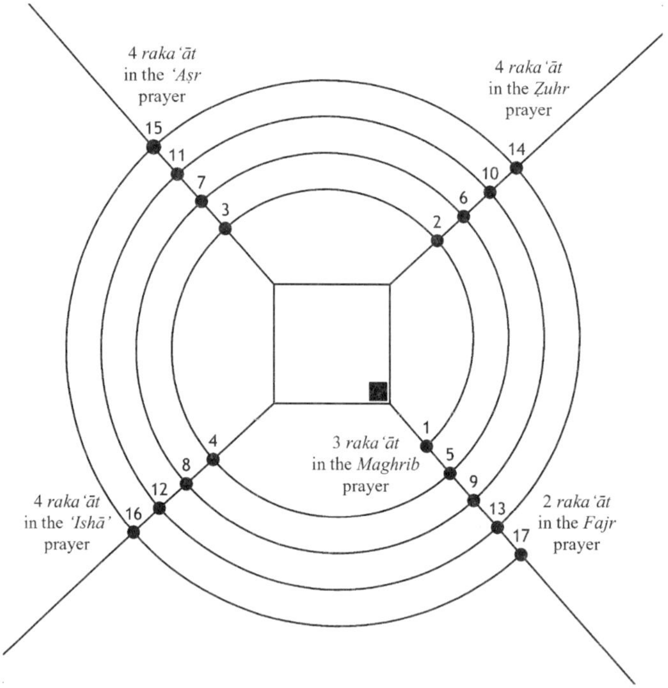

Let us start the circumambulation and leave one *rak'ah* at the beginning of every corner of the Ka'bah.

At the beginning of the 5th turn, we find that the sum of *raka'āt* at the corners of the Ka'bah is 17, and these are divided as follows:

- 5 *raka'āt* at the Black Stone. These 5 *raka'āt* are 2 and 3:
  - 2 *raka'āt*: the *Fajr* prayer (the similarity between the *Fajr* and *Maghrib* was mentioned above)
  - 3 *raka'āt*: the *Maghrib* prayer
- 4 *raka'āt* at the remaining corners and these are *Ẓuhr*, *'Aṣr* and *'Ishā'*.

This way, *Ẓuhr* and *'Aṣr* are on one side (prayers in the daylight) and *'Ishā'*, *Maghrib* and *Fajr* are on the other side (prayers when the sun is down).

If we continue with the circumambulation we find:

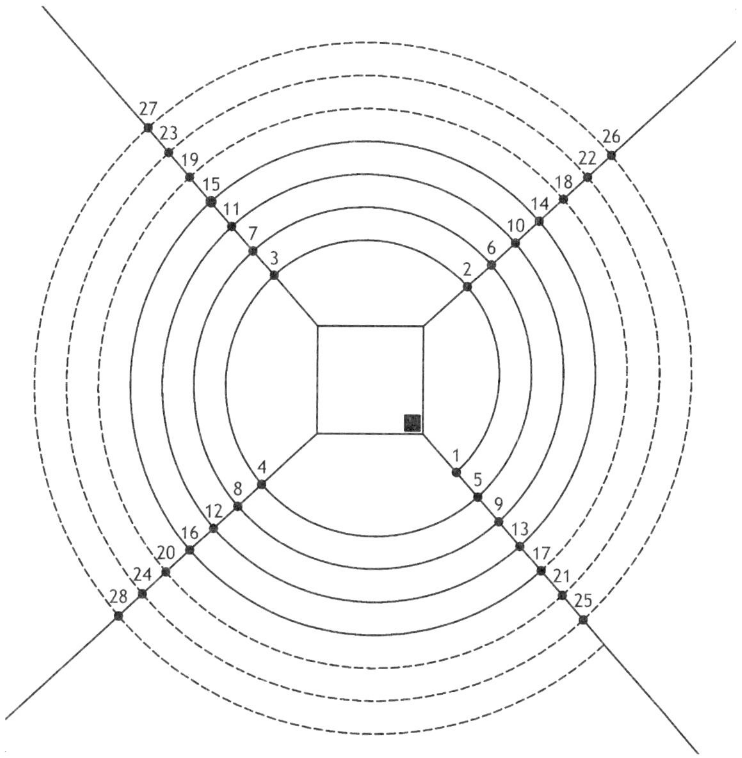

- At the end of the 7th turn, on each corner of the Ka'bah, 7 *raka'āt*, as is clear in the figure above
- 28 is the total number of *raka'āt*, 17 + 11
  - 17 is the number of obligatory *raka'āt* in one day
  - 11 is the number of the *sunnah raka'āt* in one day (as mentioned in the Ṣaḥīḥ).[51]

This shows the importance of 11 and its relation to 17.

51. Narrated by Ibn 'Umar: "I remember ten *raka'āt* of *nawāfil* from the Prophet ﷺ, two *raka'āt* before the *Ẓuhr* prayer and two after it; two *raka'āt* after *Maghrib* prayer in his house, and two *raka'āt* after *'Isha'* prayer in his house, and two *raka'āt* before the *Fajr* prayer" (Ṣaḥīḥ al-Bukhārī: 1109). Additionally, Ibn 'Umar reported that a person asked the Messenger of Allāh ﷺ about the night prayer. The Messenger of Allāh ﷺ said: "Prayer during the night should consist of pairs of *raka'āt*, but if one of you fears morning is near, he should pray one *rak'ah* which will make his prayer an odd number for him."

Let us continue with a study of Fig. 5-C1, which is the sum of the numbers on the same line in Fig. 5-C, giving this result:

- First line: 1
- Second line: 2 + 3 = 5
- Third line: 4 + 5 + 6 + 7 = 22
- Fourth line: 8 + 9 = 17

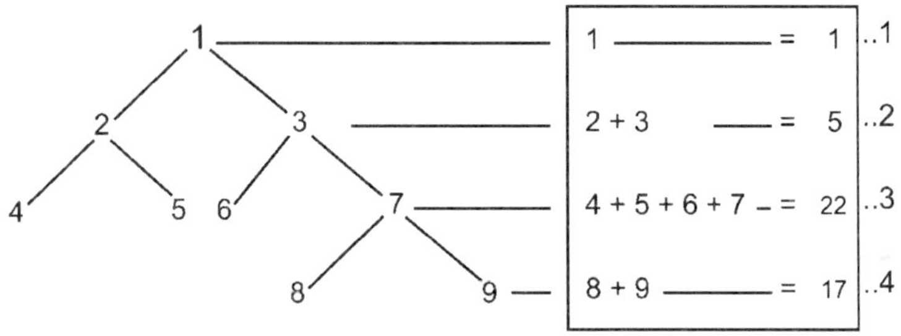

Fig. 5-C   Fig. 5-C1

We continue the study of Fig. 5-C1, but in a visual, not quantitative approach to number, taking into account that our two eyes are precious and that we should not neglect them. Allāh Most High said: *Say, Journey upon the earth and observe.* [al-'Ankabūt, XXIX: 20]

(2 + 3) = 5   (4 + 5 + 6 + 7) = 22   (8 + 9) = 17

1   2   3

Fig. 5-C2

1 + 2 = 3

(2 + 3) + (8 + 9) = (4 + 5 + 6 + 7) = 22

Fig. 5-C3

2 + 3 + 4 + 5 + 6 + 7 + 8 + 9 = 2 (22) = 4 (11)

Fig. 5-C4

$$(2 + 9) = (3 + 8) = (4 + 7) = (5 + 6) = 11$$

Fig. 5-C5

The operations shown above may seem insignificant from a quantitative approach to number. The qualitative approach that people of knowledge carry out before anything else is a visual study of number. A visual approach to numbers is fundamental to understanding them. Such an approach reveals the relationships and interactions between numbers.

In the oral teaching tradition, the operations above are stations for explanations and long studies, whereby relationships between Figures 5-C1, 5-C2, 5-C3, 5-C4, 5-C5 are understood through a visual approach to the 3×3 Wafaq square.

For example, in Fig. 5-C6, we notice that one of the most important numbers is 11. The visual approach reveals its importance.

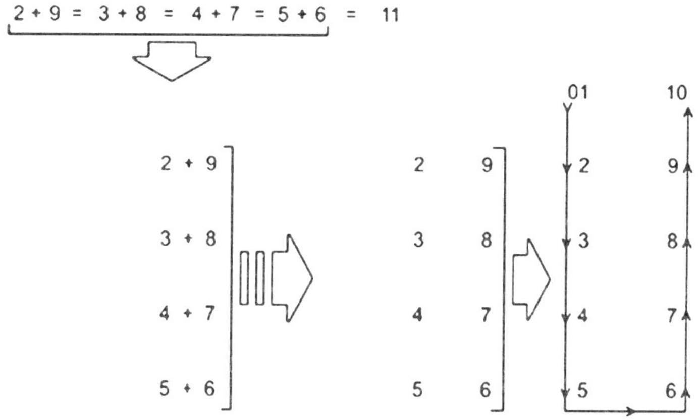

$$(2 + 9) = (3 + 8) = (4 + 7) = (5 + 6) = 11$$

$$(2 + 9) + (3 + 8) + (4 + 7) + (5 + 6) = \boxed{4\ (11) = 44}$$

Fig. 5-C6

It also reveals the number 44, which is for us:

04 and 40

If, in the 3×3 Wafaq square, we link numbers:

- 2 and 9
- 3 and 8
- 7 and 4
- 6 and 5

as they appear in Fig. 5-C6, it is interesting to note that the 4 lines, which add up to 44, are all directed to the box with number 4 (Fig. 1-D). This demonstrates the importance and reality of the visual approach to numbers.

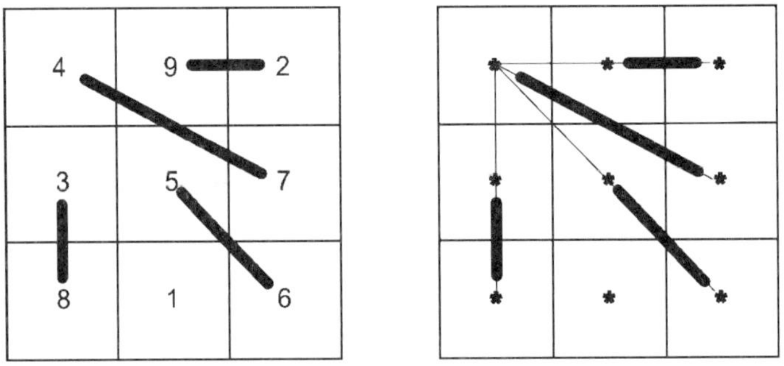

Fig. 1-D

Again, thanks to a visual approach to Fig. 5-C6, we also find that 11 is one of the most important numbers; commenting on it would be very lengthy. The operations above have demonstrated its existence, its importance, and its close relationship to the number 17.

In Fig. 5-A1, the total sum of the numbers is 17. The sum of the numbers used (5 + 3 + 2 + 1) is 11 (see Fig. 5-D and Fig. 5-E):

- 11 is essential in Islam. It is a unit of measurement, especially in the Quran and in the three formulae of faith,[52] which occur 33 times in the Quran. Their numerological values are of particular importance.[53]
- 11 is a sea of knowledge. In the present study, the relationship between 11 and 17 is clear, in particular.
- 11 is the number of *takbīrāt* in two *raka'āt*. The *rak'ah* is a unit of measurement in the canonical prayer which is the subject of this study.

52. [Editor's Footnote]: The primary formula of faith occurs in the Quran in three forms: [1] *Lā ilāha illā Llāh* [Muḥammad, XLVII: 19]; [2] *Lā ilāha illā anta* [al-Anbiyā', XXI: 87]; [3] *Lā ilāha illā huwa* [Āl 'Imrān, III: 18].

53. [Editor's Footnote]: The numerological values of the 3 formulae are respectively: 165 (11 × 15), 550 (11 × 50) and 110 (11 × 10).

The *takbīr (Allāhu akbar)* is one of the most important expressions in Islam. The *takbīr* marks the beginning of every one of the circuits of circumambulation around the Ka'bah; it is uttered while passing the Black Stone. It is the first expression in the call to prayer as well as the most important and the most characteristic, given the number of times it is repeated in the call. It is also the first phrase a Muslim hears at birth and perhaps one of the last before passing away.

In the call to prayer (pure and clear, as it was at the time of the Prophet ﷺ and the Companions), we find that the number of the letters used is as follows:

- 17 letters: 9 letters for the two formulae of faith and 8 for the rest of the words!
- 17 letters: the Arabic alphabet (28 letters) minus 11 letters.

The *takbīr* is the leading formula in the prayers. We begin the prayer with it and it leads all the movements of the prayer. The *rak'ah* is the unit that sets the rhythm of the prayer. Its name is derived from a particular posture in the prayer, which is *rukū'* (bending over in the standing position). It is interesting to note that all the prayer movements are announced, preceded and commanded by the *takbīr*, except one movement, which immediately follows the *rukū'*. Instead, *sami'a Allāhu liman ḥamidah* (Allāh hears the one who praises Him) is uttered.

What might seem to us an anomaly is in fact a sign. The human mind favors the comfort of generalization, categorization and unity of knowledge. The human mind favors the least effort. An anomaly in such a structure invites us to be alert and to pay attention.

Such an anomaly in the structure of the *takbīr* may lead to expressive, symbolic, intentional numerical sequences which are thus significant. Let us count the number of *takbīrat* in the prayers of 1 day:

- 22 *takbīrāt* in the *'Aṣr* prayer
- 17 *takbīrāt* in the *Maghrib* prayer
- 22 *takbīrāt* in the *'Ishā'* prayer
- 11 *takbīrāt* in the *Fajr* prayer
- 22 *takbīrāt* in the *Ẓuhr* prayer

Again, we find the numbers in Fig. 5-C1 and the study of their sums, namely the totals of lines (3) and (4). As for the first and second lines, they provide us with the key to the numerical sequence, which in turn provides us with numbers 11, 17, and 22 as presented below. 1 is the total of the first line; 5 is the total of the second line. This provides a sequence.

1 + 5 = 6	17 + 5 = 22
6 + 5 = 11	22 + 6 = 28
11 + 6 = 17	28 + 5 = 33, etc.

The details confirm that there is:

- takbīr which announces the beginning of the prayer. It is the *takbīr* of *iḥrām*, followed by a recitation of some phrases,[54] then the Fātiḥah and then the Quran recitation. After that
- successive movements, each followed by a *takbīr*, except after the *rukū'*, namely
- *takbīrāt*. The fifth *takbīr* of the 5 successive *takbīrāt* might be the last in the rare case of a prayer of only one *rak'ah*, or it might be the *takbīr* that is uttered before returning to the standing position (as in the beginning *rak'ah*). In this standing position, the Fātiḥah and Quran are recited, then
- successive movements requiring
- *takbīrāt* along with a compulsory seated position and a prescribed recitation, which add up to
- 11 *takbīrāt*.

To continue, we need to say: 1 *takbīr* before standing up and a recitation of the Fātiḥah, with or without the Quran recitation, and 6 successive movements accompanied by 5 *takbīrāt*. This gives us:

- *takbīrāt*, added to the previous 11, so we have
- 11 = 17; if we continue, we have to say 5 more *takbīrāt* before the prescribed sitting position.
- 17 = 22...

We thus find clearly that the 11 *takbīrāt* in two *raka'āt* can be considered a unit of measurement.

Furthermore, the numbers of the previous sequence are present in a simple reading of the great Formula of Faith.

In the famous conversion of letters to numbers, the *Ḥisāb al-Jumal*,[55] we have:

17 = اهيا (*Ahyā*), يا هو = 22 [56] (Yā Hū), 11 = هو (*Hū*)

The letters used in these three Noble Names are 4 in number:

ي، و، ه، أ

And these 4 are contained in His Name ***Yā Hū***. Why these letters and not others?

The explanation can be found in the meaning and properties of these letters. I will only mention one aspect and this is the mathematical one. In numerology, we have: أ = 1, ه = 5, و = 6, ي = 10. These numbers (1, 5, 6, 10) are the only numbers which when

54. [Editor's Footnote]: According to the Sunnah, there are some optional supplications before starting the canonical prayer as such.

55. *Ḥisāb al-Jumal* was well known at the time of Canaan, or the Phoenicians, as they are called, and used by the Aramaic people, Hebrews and Arabs. See the definition of on p. 137.

56. It was mentioned in the commentary of al-Qurṭubī on verse 40 of Surat al-Naml that the meaning of *Ahyā* is identical to the Divine Name *al-Ḥayy* (The Living).

raised to any power give a result whose final digit is 1, 5, 6, 0 respectively:[57]

$$5^3 = 12\underline{5},\ 6^3 = 21\underline{6},\ 10^3 = 100\underline{0}$$

One reason these numbers were chosen is their stability and because of their total (1 + 5 + 6 + 10 = 22) which is the numerological value of *Yā Hū*. *Yā Hū* is also 10 + 01 + 5 + 6, which is confirmed by this visual approach.

ي = 10

ا = 01

ه = 5

و = 6

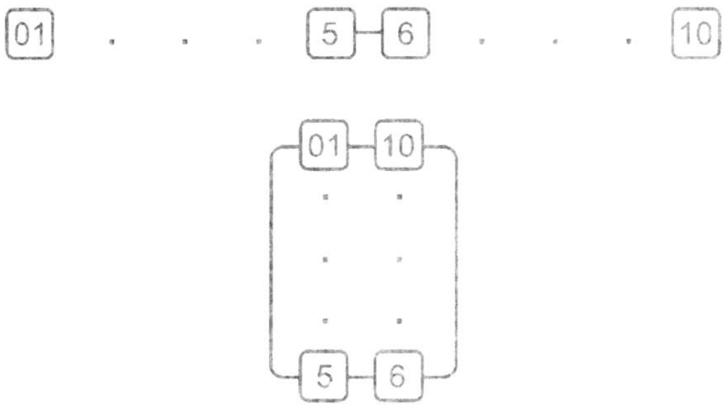

All the previous explanations (and this is something we need to remember) are readings of Figures 5-C, 5-D, 5-E.

The *takbīr* is fundamental to the 5 calls to prayer, to the 5 prayers, and in particular it commands the movements of the 17 daily *raka'āt*. The call to prayer (*adhān*) is composed of 17 distinct letters.

What could be the best number for the *takbīr* in numerological terms? It would be interesting to discover that it is 17 and it would be extraordinary if the number corresponding to the *takbīr* is something like 17 × n. The best solution is not to use a random number here, but rather use a number which has a close relationship to 17 or another number similar to it. And that n = 17 would be beyond our expectation and hope.

57. [Editor's footnote]: Likewise any number ending in 1, 5, 6 or 0, when multiplied by itself any number of times: $11^2 = 12\underline{1}$, $35^3 = 4287\underline{5}$, $216^2 = 4665\underline{6}$, $200^2 = 4000\underline{0}$.

The numerological value of *Allāhu Akbar*[58] is
$289 = 17 \times 17 = 17^2$.

There are 17 daily *raka'āt*. 8 + 9 = 17 is the number of *raka'āt* in the 5 prayers. But why 8 and 9? In fact the 17 *raka'āt* are 8 and 9:

- 8 in daylight.
- 9 after sundown.

Also 17 *raka'āt* are in fact:

- 7 *raka'āt* performed with the Fātiḥah alone.
- 10 *raka'āt* performed with the Fātiḥah and the recitation of other sections of the Quran.

7 *raka'āt* with the Fātiḥah alone? Because the Fātiḥah is famous for its 7 verses. It is commonly known to most commentators of the Quran that The Seven Oft-Repeated (*al-sab' al-mathānī*) mentioned in verse 87 of Surat al-Hijr (XV) are the 7 verses of the Fātiḥah.

All the letters of the Arabic alphabet are in the Fātiḥah, except for 7 famous letters, known as 'the letters not used in the Fātiḥah' (*sawāqiṭ al-fātiḥah*). In the correct reading of the Fātiḥah, there are 14 stressed letters and this stress (*shaddah*) is on only 7 different *abjad* letters.

10 *raka'āt* with the Fātiḥah and the recitation of other sections of the Quran? These 10 = 6 + 4 are comprised of:

- 6 *raka'āt* aloud
- 4 *raka'āt* in a soft, low voice

The 6 *raka'āt* performed aloud are among the total of 9 that are performed while the sun is down.

The 4 *raka'āt* performed in a low voice are among the total of 8 that are performed while the sun is in the sky (during daylight).

The 7 *raka'āt* with the Fātiḥah alone cannot be performed unless preceded by the 6 and 4 *raka'āt* mentioned above.[59] There is no prescribed daily prayer that is performed with the Fātiḥah alone.

There are 7 *raka'āt* performed with the Fātiḥah alone; if we consider the distribution of the 7 *raka'āt* between day and night let us try to find the best explanation by examining the number. The easiest solution might be to divide the *raka'āt* in half between day and night; in other words, to assign half to the night and half to the

58. [Editor's footnote]: The numerological value of *Allāhu Akbar* according to *Ḥisāb al-Jumal* is as follows:

ا (1) + ل (30) + ل (30) + ه (5) + ا (1) + ك (20) + ب (2) + ر (200) = 66 + 223 = 289

59. [Editor's Footnote]: These are the following: 6 = the first 2 *raka'āt* of the 3 prayers performed in the dark (*Maghrib*, *'Ishā'* and *Fajr*) and 4 = the first 2 *raka'āt* of the 2 prayers performed with the sun in the sky (*Ẓuhr* and *'Aṣr*).

day. There is no half of a *rak'ah*. Therefore, we are obliged to split them into two integers: 7 = 3 + 4.

> **This is a great sign which is manifested in the ritual walking *(sa'ī)* between *Ṣafā* and *Marwah*,[60] just as it is manifested in the declaration of faith.[61]**

Now, how can we connect 3 and 4 with 6 and 9 which represent the night or 4 and 8 which represent the daylight? I prefer to connect 3 with 6 and 9 because these 3 numbers share so many mathematical properties, especially in the form n^a. For instance:

1	2	3
4	5	6
7	8	9

$3^7 = 2187$, $2+1+8+7 = 18$, $1+8 = 9$

$6^4 = 1296$, $1+2+9+6 = 18$, $1+8 = 9$

$9^3 = 729$, $7+2+9 = 18$, $1+8 = 9$

The final result is always 9. Therefore, the way in which 3 connects with 6 and 9 is very convincing.

Similarly, 4 easily connects with the pair 4 and 8. We thus have $4 = 2^2 = 1 \times 4$ and $8 = 2^3 = 2 \times 4$. Additionally, we have a numerological connection between the words for light (*nūr*) and day (*nahār*):

light (نور) = day (نهار) = 256 = (4 × 4 × 4 × 4).

Therefore, we connect 3 with the night and 4 with the day. The ritual walking (*sa'ī*) between *Ṣafā* and *Marwah* confirms the validity of our numerical conclusion which coincides perfectly with the reality of the matter.

Therefore, 6, 4, 7 are:

- 6 *raka'āt* aloud with the Fātiḥah and Quran
- 4 *raka'āt* in a low voice with the Fātiḥah and Quran
- 7 *raka'āt* in a low voice with the Fātiḥah alone.

Similarly, we can say: There are 11 *raka'āt* in a low voice and 6 aloud. This is exactly what one can observe if one meditates upon the 5 canonical prayers as they are performed today at the Ka'bah for example, and this is exactly what the Prophet ﷺ used to do 14 centuries ago.

But what is the relationship between 11 and 6 in the structure of the prayers said aloud and in a low voice and in the sequence 1 + 5 = 6, 6 + 5 = 11, 11 + 6 = 17? We find

---

60. [Editor's footnote]: In the ritual walking between *Ṣafā* and *Marwah* there are 7 returns, and one stops 3 times at *Ṣafā* and 4 times at *Marwah*.

61. [Editor's footnote]: 3 + 4 = 7 recurs also in *lā ilāha illā Allāh* which is composed of 12 letters: 5 for the negation (*lā ilāha*) and 7 for the affirmation (*illā Allāh*), which in turn is composed of 3 letters (*illā*) and 4 letters (*Allāh*); thus 3 + 4 = 7.

more explanations in the numerological values of sacred names and words. As we mentioned before, 11 is the numerological value of *Hū* and *Hū* is one of the greatest divine names, as Rāzī said in his book *Commentary on the Divine Names*.

In the semantic study of ***Hū***, let us think of a human being who represents all of humanity. Let that human be Adam. If he says "*Hū*," who does *Hū* refer to, if we are looking for a *Hū* who is Absolute? The angels? No, because they are 'they'. Can it refer to 'X'? No, because 'X' is also 'they', as X1 implies X2 and they are objects, and *Hū* cannot be used to refer to objects.

The same applies to all creatures and existing things. We find ourselves in the final analysis faced with one truth: **there is no other *Hū* except Him** (*Huwa*) and only Allāh Most High is ***Hū***.

In this quest to find the absolute ***Hū***, we are asking who is ***Hū***? Can it be X, if every time the answer is negative? Therefore, ***Hū*** is not X, not X1, not X2, not any X, up to infinity.

Hū is Allāh (glorified and exalted be He) to Whom nothing can be compared and Who is like no one and Who is different from any creature or existing thing that you could imagine. This is the transcendence and glorification expressed by the word *Subḥāna*.

The semantic approach to the word ***Hū*** led us to *Allāh*, glorified and exalted be He, and to the word *Subḥāna*.

Let us see what a numerological study of the word ***Hū*** will lead us to. *Hū*: ه (5) + و (6) = 11.

> Contents of (11): 1 + 2 + 3 + 4 + 5 + 6 + 7 + 8 + 9 + 10 + 11 = 66.
>
> $66 = 11 \times 6 = 11 \times (11 + 1) / 2 = (11^2 + 11) / 2$.
>
> And $11^2 = 121$.

Therefore, numerically, 11 gave us simply its contents, 66, and its square value, 121.

Semantically, ***Hū*** gave us ***Allāh*** and ***Subḥān***. The numerological value of the Name of Majesty, *Allāh* = 66.

The numerological value of *Subḥān*[62] = 121 and the formula *Subḥān Allāh* is therefore 121 + 66 = 187.

187 = 11 × 17, which again confirms the close relationship between 11 and 17.

17 – 11 = 6. If we study 6 in the same way as we studied the number 5 (in Figs. 5-B and 5-A1) we will find 11 points, and the total sum of the numbers used will be 22! – as we can see in the following figures:

62. [Editor's footnote]: سبحان : س (60) + ب (2) + ح (8) + ا (1) + ن (50) = 121

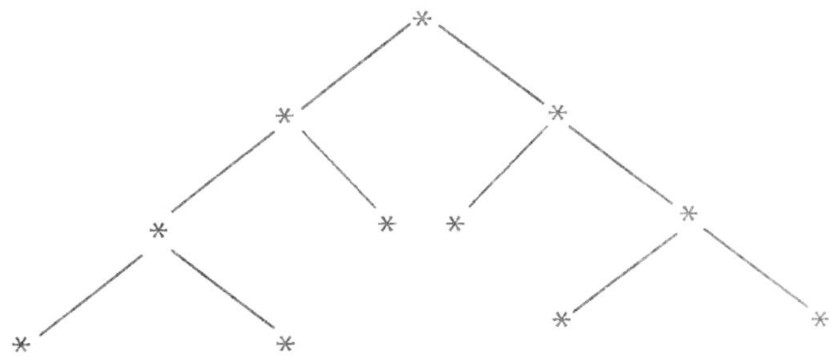

The number of points = 11

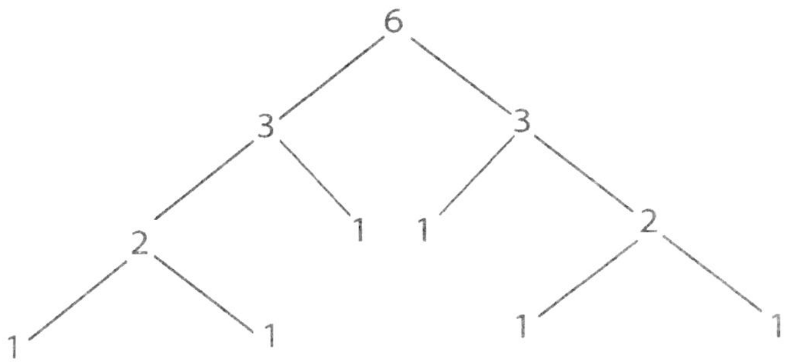

The sum of the numbers is 22

11 is the 6th prime number (1, 2, 3, 5, 7, 11).[63]

11 and 6 (as was briefly mentioned) are the keys and the foundations of the sacred geometric symbols which regulate the form of one Umayyad dome. As great and wonderful as it is, this dome is also neglected and forgotten. This is the Dome of the Chain in Jerusalem.

The Dome of the Rock is one of the world's most famous Islamic buildings. Both domes – the Dome of the Chain and the Dome of the Rock – are masterpieces of Umayyad sacred geometry and architecture. Because the Dome of the Chain is small and very close to the Dome of the Rock, astonishingly, experts consider it a tentative prototype of the Dome of the Rock. How can experts be satisfied with such a conclusion? The Dome of the Rock has an octagonal external form. The dome itself is built on a structure of $8 \times 2 = 4^2 = 16$ sides. As for the form of the Dome of the Chain, it has an external structure of 11 sides with an internal hexagon upon which the dome rests.

63. See definition of prime numbers on p. 138.

A form with 11 sides is a really unusual feature in architecture, let alone the architectural relationship between the 11-sided form and the central hexagon which bears the Dome of the Chain.

The hexagon is easy to draw with a compass, but only the masters of sacred geometry can surround the hexagon with an 11-sided form using the tools of sacred geometry (these are *regular* and *irregular* shapes). As mentioned above, 11 is the 6th prime number (1, 2, 3, 5, 7, 11).

A remarkable form such as this serves to emphasize 6 and 11 and their close relationship to 5 and 17.

$$11 - 6 = 5 \;\_\_\_\_\_\; 17 = 6 + 11$$

The contents of (11) are 11 × 6 = 66

The contents of (6) are 1 + 2 + 3 + 4 + 5 + 6 = 21

11 – 6 = 5, contents of (11) – contents of (6) = 45 if you recall.

11 and 6 are represented in the structure of the loud and silent *raka'āt*; they remind us of the sequence 1 + 5 = 6, 6 + 5 = 11, 11 + 6 = 17... They remind us of His Noble Names ***Hū*** and ***Allāh*** (may His Glory be exalted).

This building, which has great symbolism, is named the Dome of the Chain of our Master Solomon. Remember the verse of the Quran: *"Truly this community of yours is one community, and I am your Lord. So worship Me!"* [al-Anbiyā', XXI: 92]

The Dome of the Chain is one of the kinds of memorial built as reminders about the prayers prescribed to the Islamic nation, as a continuation and revival of a great, ancient, noble heritage. The dome was built as a sign and a reminder and to represent the higher knowledge that was transmitted to us through the canonical prayer.

In the Hadith of the Prophet ﷺ about his Night Journey, he informed us that Allāh, glorified and exalted be He, prescribed fifty prayers every day and night for his nation and how these were reduced five at a time, upon the advice of our Master Moses, to finally become five canonical prayers that, in recompense, equal fifty. This numerical detail is a clear sign and this is what is important to us.

I feel I have to note that the story of the Prophet's Night Journey to Jerusalem and his Ascension (*Isra' wa Mi'raj*), may seem at first a very primitive and naïve story. It also caused a lot of confusion and speculation from the moment it was reported by the Prophet ﷺ right after the journey.

Even until now, many accept the account with some resignation. Others, with more so-called 'realistic' thinking, did not want to discuss the subject, nor did they want it to be explained or commented upon, so as to avoid trouble or confusion.

Many tried to explain the matter with faith, but with very limited knowledge or understanding. The attempts yielded very 'awkward', 'strange' or naïve and simplistic results that are not higher in intelligence than that needed for people's daily obligations.

Reported in the Hadith collections of al-Bukhārī and Muslim, by one of the closest companions, our Master Anas (may Allāh be pleased with him), is a Prophetic saying about the Night Journey and Ascension. In that Hadith, Allāh informs the Prophet:

> *"... O Muḥammad, there are five prayers every day and night and each prayer is equivalent to ten, which add up to fifty prayers ...."*
>
> (Ṣaḥīḥ Muslim: 234)

This is a subtle and clear sign pointing to one of the most important forms representing numbers. Like their predecessors, the Pythagoreans received this knowledge orally. That was their great secret. It appeared in Greece after the tragic end that their community faced. After that time, this universal form representing numbers was given the famous Greek name, Tetraktys.

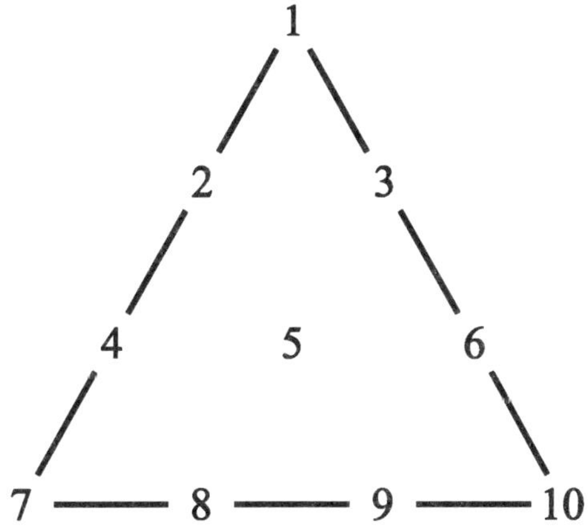

The Tetraktys

Pythagoras must have seen this form very commonly found in the architecture of Syria and Mesopotamia. We can even see it in the present day in the Amrit Temple, for example, on the Syrian Coast. The extreme simplicity of the Tetraktys, or the Triangle of the Gnostics, makes it difficult to understand. **Greatness lies in simplicity.**

This is the real distribution of numbers on the Triangle of the Gnostics, as I received it. It is possible to see this figure in 2×3 different ways, in the same way that any 3×3 square can be seen in 2^3 ways.[64]

64. [Editor's Footnote]: The 3×3 Wafaq square can be rotated to four positions relative to the viewer,

The following notes are only to provide a guarantee to confirm the perfect distribution of the numbers on the figure.

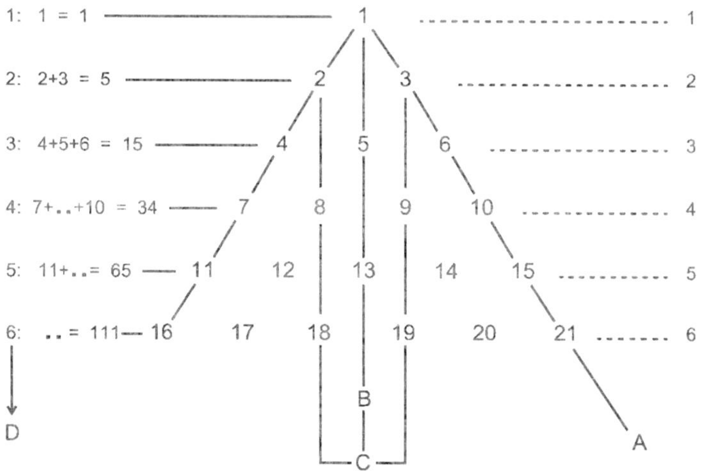

(A) Automatically gives us a number sequence (1, 3, 6, 10, 15, 21, etc.). Each row of the Tetraktys is numbered. The contents of the row number corresponds to the number in the sequence. For example: the contents of (4) = 10, the contents of (5) = 15.

(B) Gives us the number of the central box in an odd-numbered square. For example: The central number in the 3×3 Wafaq square is 5 and in the 5×5 square is 13.

(C) Gives us two of the numbers in the four central boxes in an even-numbered square.

(D) Gives us the total along any line in a Wafaq square, by simply adding the numbers in that row across to the opposite side of the figure above. For example, if we wanted to know the sum of the numbers in any line in a 4×4 Wafaq square we would take the fourth row, add up the numbers on that row: 7, 8, 9, 10, which gives us 34, and this is the sum of the numbers on any line in a 4×4 Wafaq square. If we multiply this total by the number of the row in the figure above, we will have the sum total of the whole square in question, as in our example: 4×34 = 136, which is the sum total of the whole 4×4 Wafaq square.[65]

The above notes (A), (B), (C) and (D) are only quick hints to give a guarantee, confirming the perfection of this distribution.

---

and each of these can be reflected, to make a total of 8 views. Likewise, the Triangle can be rotated to three positions relative to the viewer, and each of these can be reflected, to make a total of 6 views.

65. [Editor's footnote]: (A), (B), (C) and (D) above, which refer to the lines shown in the figure, show the close link between the Tetraktys and the Wafaq square.

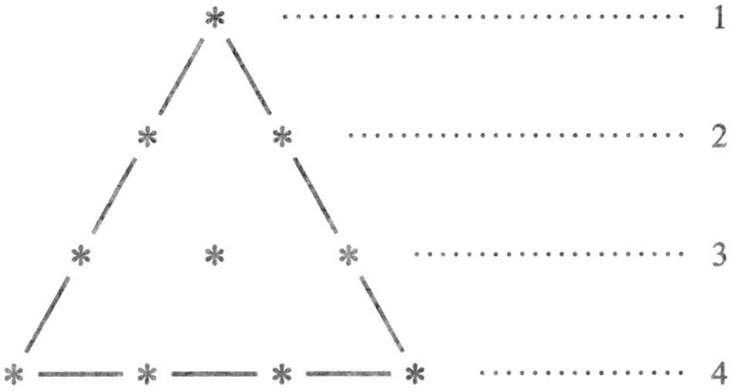

The Tetraktys

Let us go back to the Pythagorean reading of the Triangle of the Gnostics as points on four rows: it is the famous equation 1 + 2 + 3 + 4 = 10. They said that 10 is 1. It is more understandable if we look at it this way: 10 is 1 and 0; 0 + 1 = 1. **For us, 10 corresponds to 01.**

In fact, it is also 10 01

The same way that 1 is the number of unity, 10 is the number of perfection. Can there be many absolute 'perfections'? If there are, they would either be different or similar. If they are different, one of them must be more perfect than the others, which makes the others less perfect, and this is a contradiction. If they are perfectly similar then they are inevitably one. In fact, the absolute perfection is clearly one. Therefore, one is absolute perfection and it is perfect because it is absolute.

We can say that unity and perfection are inseparable. This allows us to understand: *"... O Muḥammad, there are five prayers every day and night and each prayer is equivalent to ten, which add up to fifty prayers..."*

01 prayer perfectly equals 10 prayers. Unity and perfection are inseparable and they are combined in 01 + 10 = 11, which is His glorified Name *Hū*, and can be translated in the world of names as the 'Only Perfect' or the 'Perfect One,' in other words, the One Who is Unique by His Perfection. These might be beautifully expressive phrases, but they are in fact only far echoes of the transcendent eloquence of *Hū* or 11.

01, 10 are the first and final readings of the Triangle of the Gnostics seen as points on four lines, read this way: 1 + 2 + 3 + 4 = 10.

If we apply the same approach to 10 points we will find, like our hands with their 10 fingers, 1 + 2 + 3 + 4 + 5 + 6 + 7 + 8 + 9 + 10 = 55, 5 + 5 = 10. **For us 55 is 50 and 05.**

And this is the same way that we looked at 10 and its correspondence with 01: "*O Muḥammad, there are 05 prayers every day and night, and as every 01 prayer is equivalent to 10, these add up to 50 prayers.*"

A visual approach to data is the best approach to a numerical message. Let us take another look at the 10 points of the Triangle of the Gnostics as numbers in the following manner:

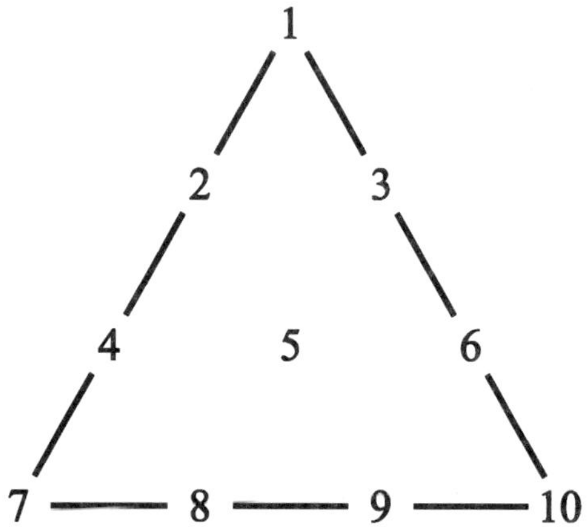

We find that there is one central number, whereas the other numbers line up around the edges of the triangle: a center and a circumference. The central number is 05 and the sum of the other numbers on the circumference is 1 + 2 + 3 + 4 + 6 + 7 + 8 + 9 + 10 = 50. The sum of the numbers on the circumference, i.e. 50, is the central 05 mirrored: "*O Muḥammad, there are 05 prayers every day and night, and as every 01 prayer equals 10, thus there are 50 prayers.*"

The close relationship between 0 and 5 can be deciphered totally through the intrinsic power of the 3×3 Wafaq square, by using very simple formulae that are highly expressive.

What I want to draw attention to is that:

(a). The Contents of (10) are **55.**
Contents of (10) = 1 + 2 + 3 + 4 + 5 + 6 + 7 + 8 + 9 + 10 = 55.
5 and 5 are like the digits of the hand, which give more explanations of the Triangle of the Gnostics.

(b). **55** is our number 11 (see above) multiplied by 05: 11 × 05 = 55.

(c). **55** is the 10th number of what is called the Fibonacci Sequence:

1,	1,	2,	3,	5,	8,	13,	21,	34,	55
1	2	3	4	5	6	7	8	9	10.

(d). **55** = $1^2 + 2^2 + 3^2 + 4^2 + 5^2$
$1^2 + 2^2 = 05$ and $50 = 3^2 + 4^2 + 5^2$

(e). It is only possible to divide **55** using the following numbers: 01, 05, 11. Their total, 01 + 05 + 11, is none other than 11 + 6 = 17.

(f). **55** is in particular the numerological value of His Name, *Mujīb* (He Who answers supplications). This is why a Muslim supplicating Allāh raises and opens both hands: 5 + 5 = 10 digits; the contents of (10) = 55 has a meaning with great symbolism. 10 + 01 = 11 is the equivalent of saying *yā*; 50 + 05 = 55 is the equivalent of saying *Mujīb*. 11 + 55 is the numerological value of *yā Mujīb* (O, Thou Who answereth supplications).

Who is the *Mujīb*? *Hū*! *Hū* = 11. Who is *Hū*? Allāh? The numerological value of the great name *Allāh* = 66 = 11 + 55 = *yā Mujīb* (O, Thou Who answereth supplications)!

We can now return with ease to take another look at Fig. 5-C. In Fig. 5-C, we have 8 lines and 9 points. We gave each point a number. A pure and inspired prescribed law confirms our tracking of signs:

3 + 2 = 5 and 9 + 8 = 17, like 17 *raka'āt* in 5 canonical prayers.

The Triangle of the Gnostics, which is closely linked to the number of prayers, gives us a strong visual confirmation.

5 is at the center. 2 and 3 on one side face 8 and 9. A real inspired prescribed law appears to confirm, explain and be explained:

- 5 prayers, 2 in daylight and 3 at night.
- 17 *raka'āt*, 8 in daylight and 9 at night.

We have a further confirmation through a simple visual and mathematical reading of the Triangle of the Gnostics.

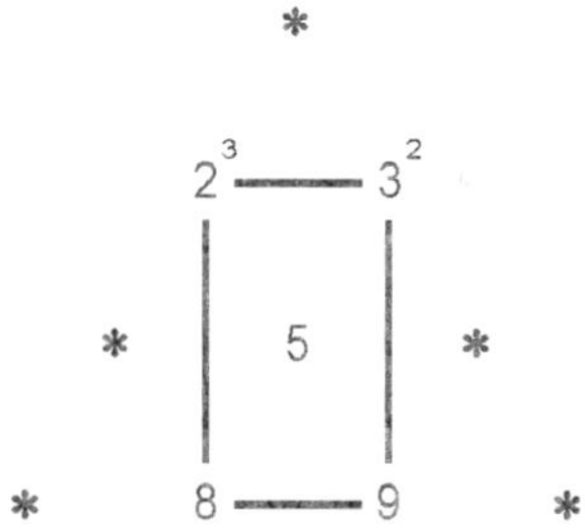

In this way, the relationship between the number of the daily prayers and the daily *raka'āt* is strongly and boldly asserted:

2 prescribed prayers in daylight	$2^3 = 8$ *raka'āt* in daylight.
3 prescribed prayers at night	$3^2 = 9$ *raka'āt* at night.

The details of the Hadith of the 05/50 prayers provide great explanations and allusions, especially the details related to the manner in which the number of prayers was reduced. Understanding this point depends on another diagram, but we consider this to be at a higher level than that of the current study.

Before we close, with great reverence, in awe of that magnificent sign about the 05/50 prayers and the related Hadith (which is in fact really far from being the folklore of idle Bedouins), let us take another look at the Triangle of the Gnostics.

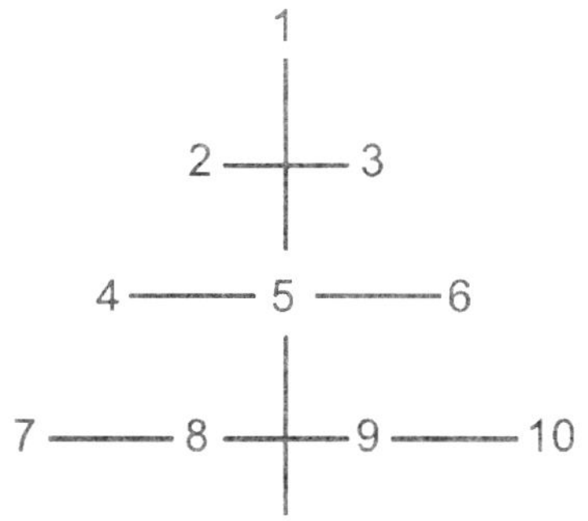

2 prayers in daylight	3 prayers at night
8 *raka'āt* in daylight	9 *raka'āt* at night
4 silent *raka'āt* in daylight with Quran recitation	6 *raka'āt* aloud at night with Quran recitation
7 *raka'āt* with Fātiḥah alone	10 *raka'āt* with the Fātiḥah and Quran recitation

We talked about the 5 prayers which are 1 of the 5 pillars of Islam. The 7 *raka'āt* performed with the Fātiḥah alone are on the side of the sun, because 4 *raka'āt* of these prayers are performed in daylight and 3 are in the night, and 4>3. The 10 *raka'āt* performed with Quran recitation are on the side of the night, because 6 *raka'āt* of these prayers are performed at night and 4 are in daylight, and 6>4.

The Triangle of the Gnostics is built on 1, 4, 10:

- 1 triangle
- 4 rows
- 10 points

The Contents of (1) + the contents of (4) + the contents of (10) = 01 + 10 + 55 = 01 + 10 + 05 + 50 = 11 + 55 = 66

$1^2 + 4^2 + 10^2 = 01 + 16 + 100 = 117 = (7 \times 17) - 2$ [the Friday *Jumu'ah* prayer has only 2 *raka'āt*!!!].[66]

117 is the number of *raka'āt* in the prescribed canonical prayers in a week.

One last glance at the Triangle of the Gnostics provides us with stronger confirmations of the number of *takbīrāt* in the prayers. 11, 17, 22, are the numbers we found through the sequence 1 + 5 = 6, 6 + 5 = 11, etc., which is confirmed by a study of the number 11, just as we did for number 5 in Figures 5-A1 and 5-C.

Let us start from the central 5, proceed towards 4, 7, then 8, 9, and lastly to 10, 6, 3, 1, 2 as in the following figure:

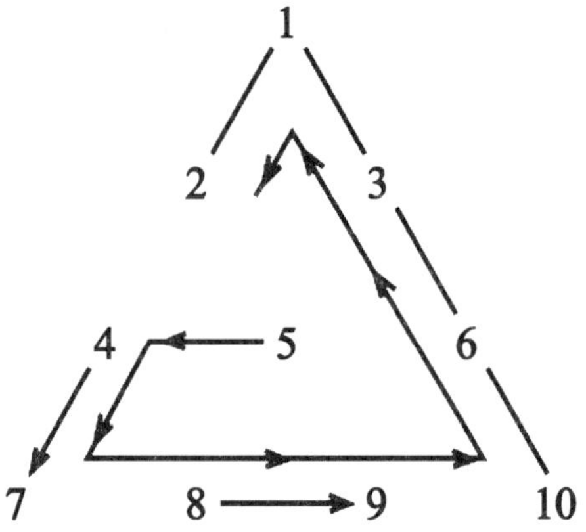

66. [Editor's Footnote]: The 4 *raka'āt* of the *Ẓuhr* prayer are replaced on Friday by 2 *raka'āt* performed in a group in the mosque.

Therefore, the Triangle of the Gnostics takes the following shape:

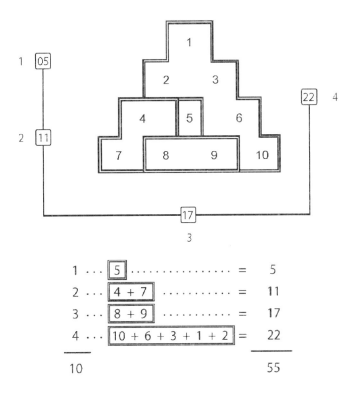

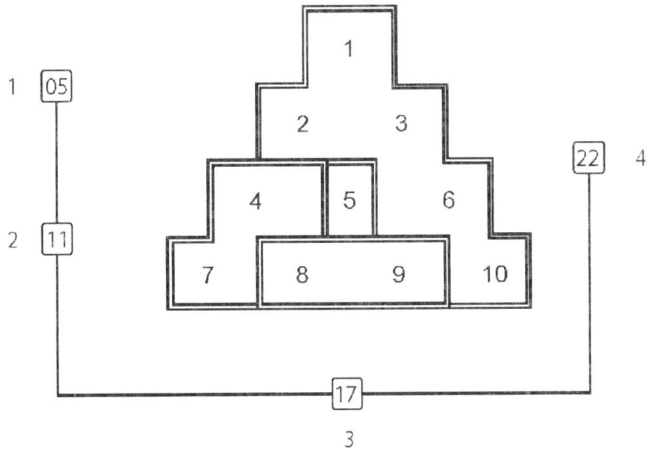

1 ...	5	=	5
2 ...	4 + 7	=	11
3 ...	8 + 9	=	17
4 ...	10 + 6 + 3 + 1 + 2	=	22
10			55

10 is the sum of the line numbers.

55 is the sum of all numbers in the Triangle of the Gnostics.

The contents of (10) = 55 = His Noble Name *Mujīb*, if you recall!

The *takbīr* in the prayers requires the utterance of *Allāhu Akbar.*

Allāhu Akbar is the phrase which characterizes the *adhān* (the call to prayer) in the 5 daily prayers...if you recall!!

The Prophet Muḥammad ﷺ said: "At the time of the call to prayer, the doors of Heaven open and supplications are answered; at the time of the *iqāmah* (the second call to prayer, given before the prayer begins) no supplication is rejected."

A numerical truth in a reading of sacred geometry as shown in Fig. 6 explains a pure and an inspired prescribed law: There were 50 prayers that were reduced to 05 prayers. 50 - 05 = 45.

45 is the number of the prayers reduced. If 05 prayers are comprised of 17 *raka'āt* ....

05 prayers of	17 *raka'āt*
45 prayers of	153 *raka'āt*

The contents of (9) = 45 = 9 × 5 ...17 × 9 = 153 = the contents of (17)

Fig. 6

Therefore, Fig. 6 is very far from acrobatically playing with numbers.

Another confirmation in another field provides a deeper conviction and a stronger belief in the unity of higher knowledge: the Major Arcana of the Tarot[67] are closely linked to the story of our Master Moses and al-Khiḍr in Surat al-Kahf.

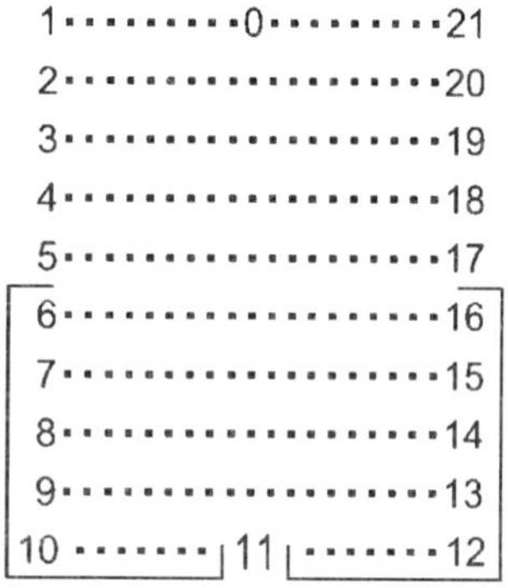

Fig. 7

Take a number, skip 5 points forward[68] and pick the 6th number as in the sequence 1 + 5 = 6. In Fig. 7, we are interested in number 5; there is no smaller number if we count five numbers before it, so we skip 5 points after it and take the 6th, which is 11. Then we skip 5 points and take the 6th, which is 17, which comes opposite to number 5. Their total is 22 as in the case of all the numbers facing each other in Fig. 7. Only 11 is facing 0. Therefore, in our quest, we found 17 *raka'āt* and 5 prayers and the structure of 11, 17, and 22 *takbīrāt*.

I am not going to dwell on the symbolism of this fact.

67. See definition of Tarot on p. 135.

68. [Editor's Footnote]: What is described here is a process of laying down the 22 cards of the Major Arcana, and moving through them in a sequence, skipping five cards and exposing every sixth card. It is significant that these cards are understood to be arranged in order, following the Major Arcana sequence, rather than shuffled to give a random result, as is the case with Tarot cartomancy.

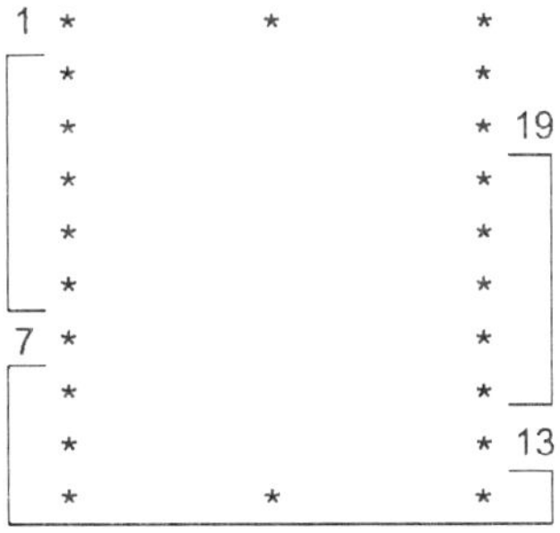

If we carry on after 17, as before, we skip 5 points and take the 6th number, then we reach 1 which is 'the Wise' in the Tarot.[69] We continue in a similar manner and we have the cards and numbers or Arcana 7, 13, then finally 19, which are the least distorted cards within the least disfigured of the copies of the Tarot, the *de Gébelin* copy.[70] It was very amusing for me to track the distortions of the cards from one copy to the other.

Certain details of the Tarot cards can be seen in some of the minarets of Damascus. Damascus is the homeland of the rose.

In 1613, a book entitled *Fama Fraternitatis Rosae Crucis* spoke of a group called the Rosicrucians, founded by a German, Christian Rosenkreutz who received in Damascus some sciences from the elite of the people of knowledge of this city in 1404 when he was 26. His father was the last of the lords of the castle of Germelshausen in the Thuringian forest. When Christian Rosenkreutz returned to Germany, he hid in a cave and transmitted his knowledge to three people. He passed away at the age of 106.

Card number 7 (the King) is so distorted! We might stop for a while in wonder at these two horses with no back limbs, stuck to this peculiar cart. But by comparing this card to classical sculptures, and pictures inspired by them - and by these I mean depictions related to Neptune, the King of the Sea - especially during the Renaissance era, when these cards were drawn, we can understand how the distortion occurred.

Card number 13 is the card of death and it is the death of the boy (in the story of Moses and al-Khiḍr).

Card number 19 is the least distorted card and it is clearly a depiction of the two

69. Refer to the definition of the Tarot on p. 135 that features Tarot images for the numbers 1, 7, 13, and 19. [Editor's comment]: The designation 'the Wise' for card 1 of the Tarot is given by the author, as well as the other names for the cards given here.

70. [Editor's footnote]: Antoine Court de Gébelin (died 1784) was a French philosopher, Protestant pastor and freemason, who published commentaries on the Tarot, suggesting esoteric meanings of the illustrations on the 22 cards of the Major Arcana, which he claimed derived from ancient Egypt.

orphans standing by the wall of the treasure (as in the story of Moses and al-Khiḍr).

If you say treasure, I would say gold, if you say gold, I would say sun. Tell me, who always uses the sun to speak of gold?

❋❋❋❋

We saw the close relationship between 5 and 17 in the canonical prayer and in the preparation of the 3×3 Wafaq square, in addition to the confirmations provided by the numbers themselves.

The total of the numbers used in Fig. 5-A1 and in the study of the number 5 is none other than 17, as in Fig. 5-D. Let us study the number 17 in the same way as we studied the number 5.

Our study of the number 5 is completely compatible with the 3×3 Wafaq square. It is also compatible with the 5 prayers, as in Fig. 5-A2. If we find the same results in our study of 17, it would be a final and absolute confirmation of the study of one of the 5 pillars of Islam - the canonical prayer - through two-dimensional sacred geometry. This is built upon numerical foundations and a set of primary, basic and grand rules, in complete harmony with the purpose and the symbolism of the 3×3 Wafaq square, given that this figure is among the noblest and purest examples of sacred geometry.

Let us study 17 in the same way as we studied 5. We separate the number 17 into its closest integer values, in other words, 8 and 9, and then divide those the same way:

- 8 is 4 and 4
- 9 is 4 and 5

We continue this way until we get to the number 1 as in the figure below:

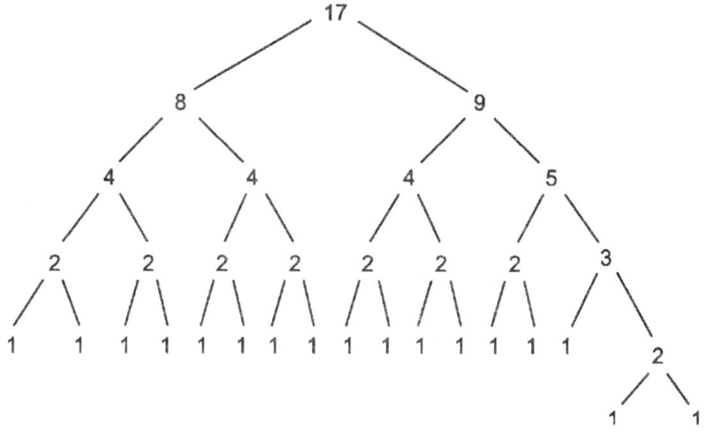

Decomposition of the number 17

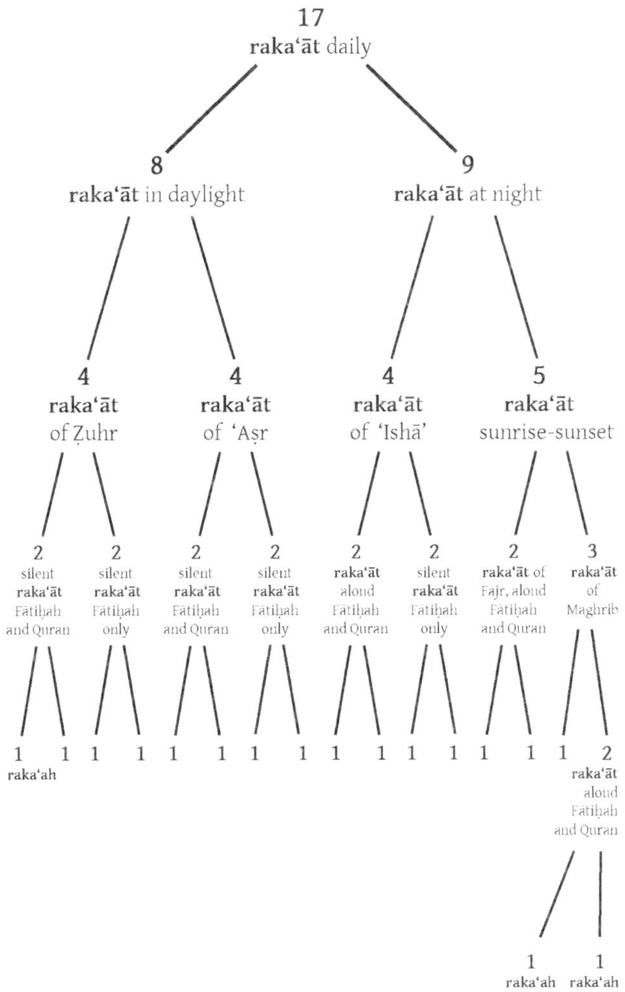

There are 5 steps in preparing the 3×3 Wafaq square. In the analysis of 5 in Fig. 5-A1, the total of the numbers is 17. In Fig. 5-A2, we have the 5 daily prayers, and a total of 17 *raka'āt*..

,..

In the Name of Allāh, the Lord of Mercy, the Giver of Mercy

...Praise be to Allāh, Who has favored us above many of His believing servants. [Surat al-Naml, XXVII: 15]

...My Lord! Inspire me to give thanks for Thy blessing wherewith Thou hast blessed me and my parents and to work righteousness pleasing to Thee, and cause me to enter, through Thy Mercy, among Thy righteous servants! [Surat al-Naml, XXVII: 19]

...This is of the Bounty of my Lord, to try me whether I will give thanks or be ungrateful. And whosoever give thanks, he gives thanks only for his own soul; and whosoever is ungrateful, truly my Lord is Self-Sufficient, Generous. [Surat al-Naml, XXVII: 40]

And who does greater wrong than one who has been reminded of the signs of his Lord, then turns away from them and forgets that which his hands have sent forth? Surely We have placed coverings over their hearts, such that they understand it not, and in their ears a deafness. Even if thou callest them to guidance, they will never be rightly guided. [Surat al-Kahf, XVIII: 57]

Definitions

The Number *Pi*

Pi (π) is a mathematical constant widely used in mathematics and physics. It is also known as 'Archimedes constant.' It is defined as the ratio of a circle's circumference to its diameter. π is approximately equal to 3.14159.

When the diameter of a circle is equal to 1, its circumference is equal to π. It is particularly used in the calculation of triangles and mathematical geometry.

π has remained a mystery. No one knows all the numbers that come after the decimal point and no one can guess how far they go.[71] This, for example, is the equivalent of π:

> 3.14159265358979323846264338327950288419716939937510582097494459
> 23078164062862089986280348253421170679821480865132823066470938446
> 09550582231725359408128481117450284102701938521105559644622948954
> 93038196442881097566593344612847564823378678316527120190914564
> 856692346034861045432664821339360726024914127372458700660631558
> 8174881520920962829254091715364367892590360011330530548820466521
> 384146951941511609433057270365759591953092186117381932611793105
> 118548...

71. [Editor's Footnote] It can be said with certainty that it has an infinite number of digits after the decimal point, and these never repeat in any pattern – so they can never all be known.

The Tetraktys or the Triangle of the Gnostics

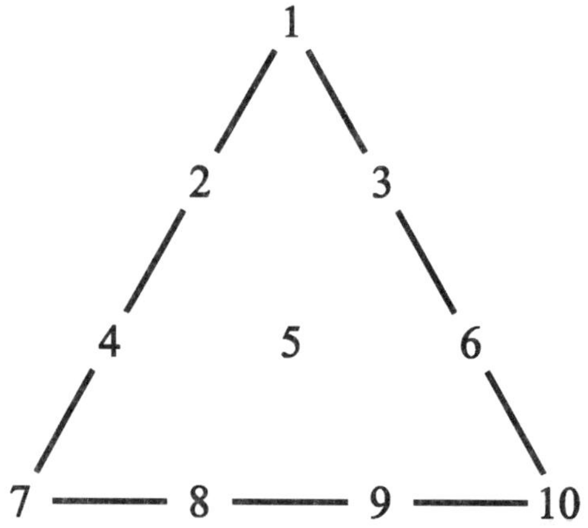

The Tetraktys is a triangular shape made of 10 points, distributed along four rows as we can see in the figure above. On the first row there is one point, on the second two points, on the third three points, on the fourth four. These four rows together lead to the number 10.

This figure bears the symbolism of the triangle and contains the number 10 and its meanings in terms of Perfection and Oneness.

The Tetraktys is considered one of the secret and mysterious symbols in sacred geometry and was considered the greatest secret of the Pythagoreans because it contains the highest levels of knowledge, to the point that revealing the secret was enough for one to lose one's life.

The Tarot

The Tarot first appeared in Europe in the 15th century. It is comprised of a deck of 78 cards, of which 56 are called the *Minor Arcana*[72] and 22 are called the *Major Arcana.*

Tarot cards are widely used in most of Europe, mainly for fortune telling.

Below are cards 1, 7, 13 and 19 as examples:

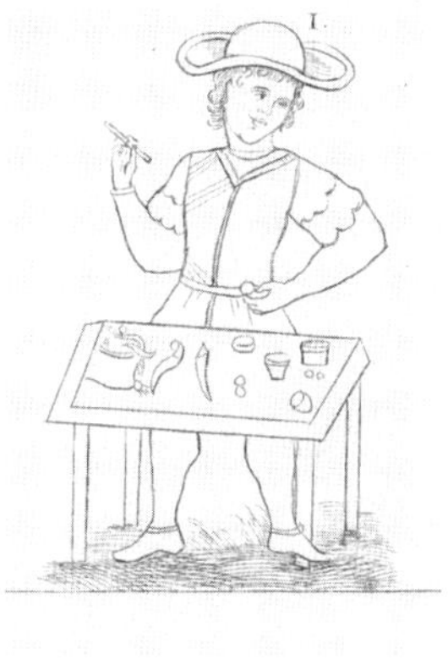

1
The Wise

7
The King

13
Death

19
The orphan boys
and the wall

72. [Editor's Footnote]: There are four suits in the *Minor Arcana*, each with 14 cards, consisting of 10 'pip' cards [1 (Ace) to 10] and 4 'court' cards [Page, Knight, Queen and King]. 4 x 14 = 56 cards. The Topkapi Museum in Istanbul has three partial decks of Islamic playing cards, called Mamluk cards, dating from the 15th century. These employ an identical system, and are considered to be the prototype of modern playing cards. Documents recovered from the Cairo Geniza include a Tarot card, apparently from the 12th or 13th century, and therefore Ayyūbid or Fāṭimid.

The *Gestalt* Theory

The *Gestalt* theory was founded by the Austro-Hungarian psychologist, Max Wertheimer. According to this theory, we perceive parts of objects and, on the basis of these, recognize the form of the whole.

No separate part in a given figure has a meaning by itself. Its meaning is derived when seen together with the whole. The whole also gives meaning to the parts.

For example, suppose we have a number of components: one large circle, two small circles, and two lines. Each component, considered alone, would have a separate meaning from the others, and no interrelationships. But if they are arranged in a certain manner, they will form a simple picture of a face. These parts together give meaning to the picture of the face we can see below.

Ḥisāb al-Jumal or the *Abjad* System

The *Abjad* system assigns a numerical value to each letter of the Arabic alphabet. It starts with 1 for the letter *alif* (أ) and ends with 1000 for the letter *ghayn* (غ). The order of *Abjad* letters and their values are listed in the following chart:

أ	1	ب	2	ج	3	د	4	ه	5	و	6	ز	7	ح	8	ط	9
ي	10	ك	20	ل	30	م	40	ن	50	س	60	ع	70	ف	80	ص	90
ق	100	ر	200	ش	300	ت	400	ث	500	خ	600	ذ	700	ض	800	ظ	900
غ	1000																

This numerological system was used in the Semitic languages, in ancient India, by the Hebrews and in the Aramaic alphabet.

After the revelation of the Quran, the expansion of Islam and of the Islamic empire, there was a need for calculation and the *Ḥisāb al-Jumal* was applied to the Arabic language and was used for a long time. Arabs used it in their sciences, trade, astronomical tables, and for calculating weights, as well as for marking the dates of their battles, deaths and buildings. For instance, Abū Rayhān al-Bīrūnī, who lived in the 4th and 5th century AH, used *Ḥisāb al-Jumal* extensively in his book *Al-Qānūn*. Note: *Ḥisāb al-Jumal* is the system used in the present book.

THE GOLDEN RATIO: Φ

Phi, or the Golden Number, or the Divine Ratio, is the eternal secret of the science of beauty and the proportion by which creativity is guided. All these terms became known after Leonardo Fibonacci put forward the famous sequence that is named after him.

The numbers of the Fibonacci sequence are as follows: 1, 1, 2, 3, 5, 8, 13, 21, 34, 55, 89, 144.... Each number is the sum of the two preceding numbers. Additionally, each number divided by the preceding one gradually approaches 1.618, or the number φ.[73]

THE PRIME NUMBERS

These are natural numbers that have no positive divisors other than 1 and the number itself.

A composite number is a natural number that is greater than 1 but not prime. For instance:

- 5 is a prime number because it can only be divided by itself and 1.
- 6 is a composite number because it can be divided by itself and 1, but also by 2 and 3.

Together, prime numbers constitute a group (1, 2, 3, 5, 7, 11, ...).

73. [Editor's Footnote]: The Golden Ratio is an irrational number and as a unit ratio its value is $(1+\sqrt{5})/2$. The 'Fibonacci' sequence of whole numbers, which increasingly accurately approximates the Golden Ratio, was first documented in 6th century India.